MASTERING

CHEMISTRY

D0755303

B349

MACMILLAN MASTER SERIES

MASTERING
CHEMISTRY

P. CRITCHLOW

M

First published 1982 by
THE MACMILLAN PRESS LTD
London and Basingstoke
Companies and representatives through-
out the world

Typeset in Great Britain by
REPRODUCTION DRAWINGS LTD
Sutton, Surrey

Printed in Hong Kong

ISBN 0 333 31294 5 (hard cover)
    0 333 30447 0 (paper cover – home edition)
    0 333 31064 0 (paper cover – export edition)

This book is also available under the title
Basic Chemistry
published by Macmillan Education.

DEDICATION

To Jane, Iain and Alec

CONTENTS

CONTENTS

CONTENTS

PREFACE

This book has been written with the dual aim of taking a reader with no previous grounding in science to a familiar understanding of chemistry, whilst at the same time providing a revision text for the more experienced student.

The content is suitable for those on a course leading to basic school or college examinations.

Anything more than the most basic arithmetic process is limited to the final part of the book, in the belief that the mole concept is more easily grasped when seen as one integrated block. Furthermore, readers seeking an outline knowledge of chemistry could miss this part altogether and still achieve a good working understanding.

The naming system adopted here is that suggested for non-advanced work by the Association for Science Education in their booklet *Chemical Nomenclature, Symbols and Terminology* (1979).

My thanks to parents for sustenance whilst working towards my deadline, to my editor for tolerance when working beyond my deadline and to friends for their helpful advice.

Peter Critchlow

PART I
FUNDAMENTALS

This part attempts to sort from the historical jumble of scientific discovery just those ideas and experiments upon which modern chemistry is based.

CHAPTER 1

VERY FIRST PRINCIPLES

1.1 ELECTRIC CHARGE

A rod of plastic, a pen for example, that has been rubbed with cloth attracts small pieces of paper or hair. The pen has been 'charged with electricity'. Two pens 'charged' in this way will repel one another, as the experiment in Fig. 1.1 shows. The same results are obtained when two glass rods are used in place of the two plastic pens.

Identical substances rubbed in the same way must gain the same electric charge. The experiment therefore suggests that

Like electrical charges repel each other

If a piece of rubbed glass is held near the suspended charged pen, the pen moves towards the glass — it is attracted. Had the glass and the pen possessed identical charges, they would have repelled one another. The glass rod must therefore be differently charged. Experiments of this kind show that two types of electrical charge exist, and that these

Opposite electrical charges attract each other

The two types of charge are described as positive (+) and negative (−). In the above examples, the rubbed glass is said to be positively charged and the plastic pen therefore negative.

No-one can explain *why* electrical charges behave as they do, but nevertheless a good starting point in understanding chemical behaviour is this assumption that like charges do repel and unlike charges do attract.

1.2 ENERGY

The faster an object moves, the more energy it has. This form of energy, which an object has because of its motion, is called *kinetic energy*.

In contrast, two oppositely charged objects held firmly apart possess

fig **1.1** *the suspended pen turns away as the other charged pen is brought close*

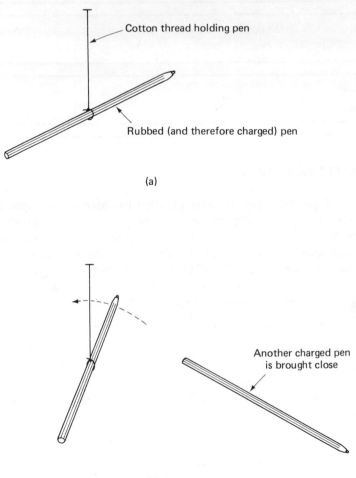

Cotton thread holding pen

Rubbed (and therefore charged) pen

(a)

Another charged pen
is brought close

(b)

a store of energy. Their opposite charges cause the objects to be attracted (see Fig. 1.2). If released, they shoot towards one another and collide. Before their release, this energy was already there, but stored as *potential*

fig **1.2** *attraction of opposite charges*

Positively charged
object

Negatively charged
object

+ ⟶ Electrical ⟵ −
attraction

energy. Potential energy is the stored energy possessed by an object experiencing a pull or a push. The pull or push may be caused by electricity, as in the above example, or by magnetism, gravity or even a spring (Fig. 1.3).

On releasing the trolley shown in Fig. 1.3, its potential energy is converted first to kinetic energy as it moves towards the wall pulled by the spring, and then to heat and sound as it hits the wall.

fig 1.3 *this trolley possesses potential energy*

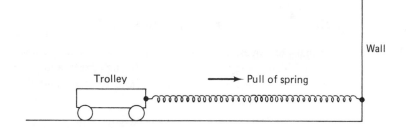

Now look at Fig. 1.4, which considers a weight on a shelf above the ground.

fig 1.4 *potential energy of a weight*

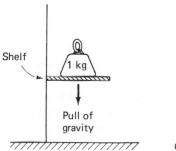
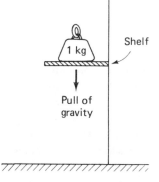

This weight possesses potential energy. If it falls from the shelf, its potential energy is converted first to kinetic energy (it moves faster and faster towards the ground, pulled by gravity), and then to collision energy of heat and sound (it hits the ground).

This weight possesses more potential energy because it is further from the ground. If it falls from the shelf, more energy is released (it will be moving faster by the time it hits the ground).

If we wish to separate two objects which are being pulled together by, for example, a rubber band, we have to put in energy. This energy that we provide becomes stored as potential energy. On releasing the objects, they shoot closer together and the potential energy is released.

A most important property of energy is that it is indestructible. It can be converted from one form into another: for example, heat energy can be converted into light, sound, kinetic or potential energy. But it can never be destroyed.

In chemistry, energy is most commonly transferred in the form of *light* or *heat* or *electricity*.

(a) Light
Light can have many energies, each one seen by us as a different colour (Fig. 1.5). Light of high energy is blue, going through green and yellow to red for light of lower energy. Light can radiate through, and therefore transfer energy across, empty space.

fig 1.5 *energies of light*

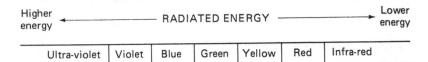

Light of energies below red is invisible, but is still there as *infra-red* radiation. Infra-red is the means by which heat from the Sun is radiated to us through space, and it is sometimes known as *radiant heat*; it is the direct heat that can be felt coming from a fire.

Energies above blue are the *ultra-violet* radiations, which are again invisible. These are also present in the Sun's radiation and are responsible for sunburn, the fading of many dyes in material and the production of vitamin D in our skin.

(b) Heat
When we say that a substance is hot, this means that its particles are vibrating or moving. When infra-red radiation strikes a substance, its energy causes the particles of the substance to vibrate more and more, thus making it hotter. When heat is passed on from a cooking ring to a frying pan, and from the frying pan to a sausage, it is really increased vibration and movement of the constituent minute particles of the cooker, pan and sausage that are passed on.

(c) Electrical energy

An electric current in a metal wire is pictured as a drift of minute negatively charged particles through the wire. The particles are called *electrons*. They will be met in nearly every aspect of chemical behaviour, since we believe that they are fundamental to all matter.

An electrical cell (often wrongly called a battery) can be pictured pushing electrons from its negative terminal into one end of the wire whilst pulling electrons from the other end into the positive terminal (Fig. 1.6). The greater the voltage, the greater this push–pull effect. When electrons are forced in this way through a particularly thin piece of wire in a light bulb, electrical energy is used up, being converted into heat and light.

fig 1.6 *a way of visualising electric current*

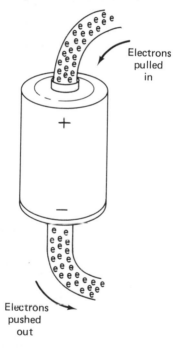

Electrons pulled in

Electrons pushed out

SUMMARY OF CHAPTER 1

Like electric charges repel, unlike charges attract.

Energy cannot be destroyed, but it can be converted from one form to another.

Stored energy is potential energy, movement energy is kinetic energy.

Energy can be radiated from one substance to another. High energies are radiated in the form of ultra-violet, with visible and then infra-red light corresponding to lower energies.

The hotness of a substance is determined by the vibration and movement of its constituent particles.

An electric current in a wire is a drift of tiny negatively charged particles called electrons.

CHAPTER 2

STRUCTURE OF THE ATOM

2.1 ELEMENTS

Common salt, sodium chloride, contains the substances sodium and chlorine. It is possible to break sodium chloride into shiny, inflammable sodium metal and green, choking chlorine gas. Similarly rust, iron oxide, can be broken down into the metal iron and oxygen gas. Copper sulphate can be split into pure samples of copper, sulphur and oxygen. The methods for splitting such chemicals into their constituent parts are varied and, as will be seen later, often require considerable quantities of energy, but nevertheless such breakdowns are *possible*.

In contrast, certain other chemicals are already single substances. For example, whatever experiment or quantity of energy is applied to a lump of copper, it cannot be broken down into simpler substances, because it is already a single, simple substance or *chemical element*. Similarly oxygen, iron, sodium, chlorine, hydrogen and gold are examples of elements. They cannot be broken down into simpler substances, since each of them is already one single substance.

2.2 ATOMS

Imagine dividing a block of an element, for example copper, into smaller and smaller pieces. Chemistry seems to be most reasonably explained if we assume that, rather than being able to continue dividing the block an infinite number of times, we would ultimately reach one single particle or *atom* of copper.

The idea that a given element is composed of tiny, identical individual atoms is fundamental to our interpretation of chemical behaviour, even though the unimaginably small size of one atom means that in practice we could never see or handle just one single atom. A small gold ring would contain about

ten	thousand	million	million	million	or	10^{22}	atoms
10	000	000 000	000 000	000 000			

of gold. One person counting out each atom from such a ring, counting as fast as possible and stopping neither for sleep nor food would require at least one-hundred-and-fifty million million years to complete the task.

2.3 INSIDE THE ATOM

Even though too small to be seen, the structure of an atom can be guessed from the results of a number of experiments. Conclusions drawn from these experiments lay foundations for the following chapters, but the details given in (a) and (b) below could well be omitted by the first-time reader.

(a) Gas discharge experiments

These were carried out by Goldstein and others in the late nineteenth century. A low concentration of a gas such as hydrogen, oxygen or nitrogen was sealed into a long glass tube. Perforated metal plates inside the tube (these are called electrodes) were subjected to a high electrical voltage in order to make one plate highly positive and the other highly negative (Fig. 2.1). This arrangement should be seen simply as a means of supplying electrical energy to the gas in the tube; an energy which in fact is sufficient to break up the atoms.

Goldstein detected a beam of positively charged particles which were attracted through the perforations of the negative electrode. The mass of these particles was later found to depend upon which gas was in the tube. Through the positive electrode shot a beam of negatively charged particles of exceptionally low mass, but always of the same mass whatever gas was in the tube.

fig 2.1 *subjecting a gas to high-voltage electricity*

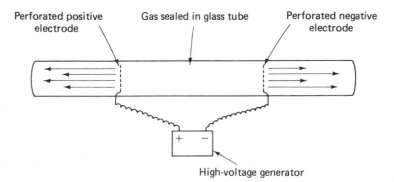

Perforated positive electrode

Gas sealed in glass tube

Perforated negative electrode

High-voltage generator

This experiment suggested that atoms of these gases contain the same light negatively charged particles (now called *electrons*) and more massive positively charged lumps whose actual mass was dependent on the nature of the element.

(b) Alpha-particle bombardment experiment
This was first performed by Geiger, Marsden and Rutherford at the beginning of the twentieth century. Many naturally radioactive elements such as uranium or radium spontaneously fire out small dense positively charged particles called *alpha particles*. A beam of such particles was directed at a thin sheet of gold leaf (Fig. 2.2). The results of this experiment

fig 2.2 *alpha particles fired at gold leaf*

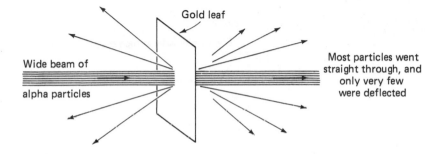

have been explained by assuming that most of the gold atom is nothing but empty space (this explains why most alpha particles went straight through) with virtually all the mass of the atom concentrated into an exceptionally small dense region which Rutherford called the *nucleus* (more rare collisions with these nuclei would explain why the occasional particle was deflected). Since the beam of particles was positive, it seems likely, when like charges repel, that these nuclei should also be positive.

These experiments begin to build a picture of the atom. Most of its mass is held in a very dense positively charged lump called the *nucleus*, surrounded by light negatively charged particles called *electrons*. The electrons are spread thinly in the space surrounding the nucleus (Fig. 2.3).

In order to explain why these negative electrons avoid being dragged into the positive nucleus (opposite charges attract), it was suggested that the electrons must be orbiting the nucleus at high speeds (Fig. 2.4).

More recently it has been shown that three major particles can be smashed from atoms (see Table 2.1). Any atom comprises a dense nucleus, positively charged and holding nearly all the atom's mass. It contains

fig 2.3 *the suggested arrangement of nucleus and electrons*

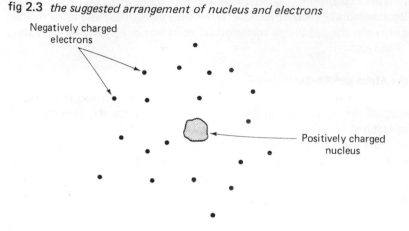

Negatively charged electrons

Positively charged nucleus

fig 2.4 *the movement of an electron in an atom*

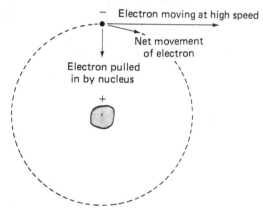

Electron moving at high speed

Net movement of electron

Electron pulled in by nucleus

electrically neutral neutrons and positive protons held very strongly together in a tiny volume.*

In the empty space surrounding the nucleus there are sufficient electrons to balance its positive charge. These electrons orbit the nucleus at high speed, and for this reason do not collapse into the nucleus.

2.4 ARRANGEMENT OF THE ELECTRONS

An electron, constantly attracted towards the positive charge of the

*Positive protons are held together in the nucleus by incredibly strong *nuclear forces.* However, nuclear energy is never released in a chemical reaction; such reactions involve only outer electrons.

Table 2.1 *the three main atomic particles*

	Mass	Charge
Protons	1 unit	+ 1 unit
Neutrons	1 unit	0
Electrons	1/1850 unit	− 1 unit

nucleus, *needs* energy if it is ever to move further from the nucleus. Heat is a form of energy, and when atoms are heated many electrons do in fact move further away from their nuclei to positions of greater potential energy.

If these electrons then jump back to their original positions all the potential energy that they gained will be given out again. In practice this energy is often given out as visible light (Fig. 2.5).

This is the process which causes very hot materials to glow 'red hot'. The heat energy excites electrons further from their nuclei. As the electrons continuously jump back, the energy is given out again, but now in the form of red light. Electrical energy, rather than heat, can be given to gases sealed into a glass tube (see Fig. 2.1). Here again, light energy is frequently given out as 'excited' electrons jump back, causing the tube to glow.

It has been noted that the light given out in such experiments is always composed of very specific light energies. For example the electrons of the element sodium, as they jump back, give out only one strong energy corresponding to yellow light (the yellow of sodium-vapour street lights).

fig 2.5 *heat can cause light energy to be given out*

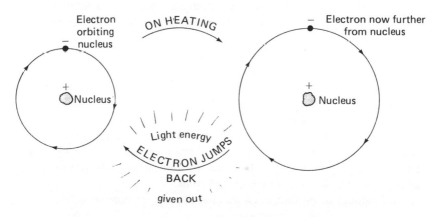

In the same way, the element strontium gives out yellow and red light. This observation suggests that the electrons themselves must exist at certain precise, specific energies, and cannot orbit the nucleus with just any energy.

We say that the electrons can orbit only within certain specific *energy levels,* the higher energy levels being at greater distances from the nucleus (Fig. 2.6).

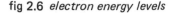

fig **2.6** *electron energy levels*

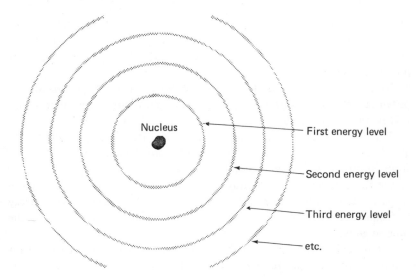

Electrical attraction would tend to pull electrons into the first energy level, closest to the nucleus, but there must be a limit to the number of electrons that can be accommodated here. Since we cannot see electrons, the electron arrangement of an atom can only be guessed at, and the guess shown in Fig. 2.7 is currently accepted as giving the best explanation of chemical behaviour.

2.5 WHY ATOMS DIFFER

We are now in a position to apply the atomic picture just described to the atoms of specific elements.

Over one-hundred different elements are known, and the atoms of each have a specific number of protons. For example, all atoms of hydrogen have just one proton each, all atoms of carbon have six protons, and all atoms of copper have twenty-nine protons.

The number of protons in an atom is the *atomic number* of the atom.

fig 2.7 *arrangement of electrons in an atom*

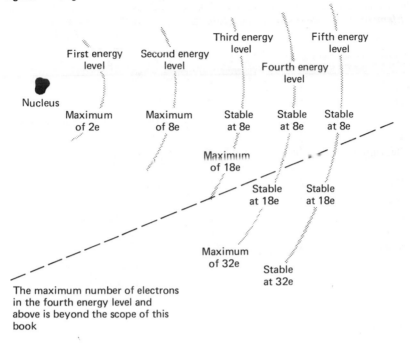

The maximum number of electrons in the fourth energy level and above is beyond the scope of this book

Thus the atomic number of hydrogen is 1, of carbon 6 and of copper 29. Symbols are given to represent each atom. One atom of hydrogen is represented as H, one atom of carbon as C and one atom of copper as Cu. Occasionally the atomic number is written just next to the symbol, as shown below:

$$_1H \qquad _6C \qquad _{29}Cu$$

Since elements, and therefore the atoms of which they are composed, are never found naturally with huge electrical charges (we would never expect to pick up a kitchen pan to find it with a twenty-thousand volt charge, for example), it must be that every proton in an atom is balanced by an orbiting electron. Hence we can say that

Number of protons (atomic number) = number of orbiting electrons

Table 2.2 shows the arrangement of electrons in the atoms of atomic numbers 1 (hydrogen) to 20 (calcium).

2.6 THE PERIODIC TABLE

Table 2.2 of electronic structures shows a most important pattern as

Table 2.2 *electronic structures of elements $_1H$ to $_{20}Ca$*

Element	Symbol	Atomic number	Electron arrangement in energy levels			
			1	2	3	4
Hydrogen	H	1	1e			
Helium	He	2	2e			
			(first energy level now full)			
Lithium	Li	3	2e	1e		
Beryllium	Be	4	2e	2e		
Boron	B	5	2e	3e		
Carbon	C	6	2e	4e		
Nitrogen	N	7	2e	5e		
Oxygen	O	8	2e	6e		
Fluorine	F	9	2e	7e		
Neon	Ne	10	2e	8e		
			(second energy level now full)			
Sodium	Na	11	2e	8e	1e	
Magnesium	Mg	12	2e	8e	2e	
Aluminium	Al	13	2e	8e	3e	
Silicon	Si	14	2e	8e	4e	
Phosphorus	P	15	2e	8e	5e	
Sulphur	S	16	2e	8e	6e	
Chlorine	Cl	17	2e	8e	7e	
Argon	Ar	18	2e	8e	8e	
			(third energy level now stable at 8e, but it can later fill up to 18e)			
Potassium	K	19	2e	8e	8e	1e
Calcium	Ca	20	2e	8e	8e	2e

soon as the *properties* of these elements are also considered.

Lithium (third in the list) is a shiny, soft metal which attacks water to form hydrogen gas and burns in air with a brilliant flame. Eight elements further on is sodium, a shiny, soft metal which attacks water to form hydrogen gas and burns in air with a brilliant flame. Eight elements further on again is potassium, with exactly these same properties.

The behaviour of helium (atomic number 2) is almost identical to that of neon, eight elements further down, and to that of argon, another eight elements down.

Properties seem to repeat after definite intervals in the list. If, instead of writing the list downwards, it is written across the page, and if every so often a new line is begun, it is possible (by carefully selecting where to start new lines) to produce a table in which elements in each vertical column have very similar properties. For example, we would make sure that lithium, sodium and potassium came in the same column; likewise helium, neon and argon.

Such an arrangement is called the Periodic Table of Elements. The positions of the first twenty elements (together with a few others) in this table are given in Table 2.3. From this table we would expect, for example, fluorine, chlorine, bromine and iodine to behave in a similar manner. Indeed, each of these elements smell similar, burn with metals to form similar solids, react with hydrogen to form similar gases, and so on.

The vertical columns are known as *groups*, which are numbered as shown in Table 2.3. The properties of elements within any one of these eight groups do in fact show a close relationship.

Two facts emerge from this.

(i) Elements in the same group have very similar properties.
(ii) It can be seen from Table 2.3 that elements of the same group have the same number of outer electrons.*

Combining facts (i) and (ii), it seems that

The properties of elements depend on their number of outer electrons

That is to say that lithium, sodium and potassium behave in a similar way because they each have just one outer electron; carbon and silicon are similar because they each have four outer electrons, and fluorine, chlorine, bromine and iodine have closely related properties because they each have seven electrons in their outer energy levels.

Given that rubidium, a rare element, has one outer electron, we would predict for it properties similar to those of lithium, sodium and potassium. In fact rubidium is in group 1 of the Periodic Table, placed immediately below potassium.

Astatine, another rare element, has a complex electronic structure but just seven electrons in its outermost energy level. We would therefore place it in group 7 of the Table, predicting that it should exhibit properties similar to the other members of that group.

Conversely, if the position of an element in the Table is known, its properties can again be predicted. For example, in the year 1869

---

*With the exception of helium, where the first energy level is not able to hold eight electrons.

Table 2.3   the Periodic Table showing the position of the first twenty elements. The electronic arrangement of each atom is shown below its symbol, although the 'e' for electron has been omitted

Group 1	Group 2	Group 3	Group 4	Group 5	Group 6	Group 7	Group 0
							Helium $_2$He 2
Lithium $_3$Li 2, 1	Beryllium $_4$Be 2, 2	Boron $_5$B 2, 3	Carbon $_6$C 2, 4	Nitrogen $_7$N 2, 5	Oxygen $_8$O 2, 6	Fluorine $_9$F 2, 7	Neon $_{10}$Ne 2, 8
Sodium $_{11}$Na 2, 8, 1	Magnesium $_{12}$Mg 2, 8, 2	Aluminium $_{13}$Al 2, 8, 3	Silicon $_{14}$Si 2, 8, 4	Phosphorus $_{15}$P 2, 8, 5	Sulphur $_{16}$S 2, 8, 6	Chlorine $_{17}$Cl 2, 8, 7	Argon $_{18}$Ar 2, 8, 8
Potassium $_{19}$K 2, 8, 8, 1	Calcium $_{20}$Ca 2, 8, 8, 2					Bromine $_{35}$Br	Krypton $_{36}$Kr
						Iodine $_{53}$I	Xenon $_{54}$Xe

Mendeleef made precisely accurate predictions of the behaviour of the element germanium (found below silicon in group 4) when the element itself was not discovered until 1871.

2.7 GROUP 0

At first sight, helium, neon, argon, krypton and xenon appear to be the most boring collection of elements. They have one property in common: a total unwillingness to do anything. They are gases that will react with virtually no other substance, and for this reason are called the *inert gases*.

Since the atoms of these gases show no tendency to change, they must be completely stable. They each have eight electrons in the outer energy level (with the exception of helium). Hence we can say that

Atoms with eight outer electrons (or two for the first
energy level) are completely stable

Combining this statement with a general observation that anything seems to show a tendency to become more stable, we are now in a position to understand why, for example, hydrogen and oxygen gases form an explosive mixture, why petrol burns and why dangerous sodium metal reacts so violently with poisonous chlorine gas to form harmless table salt.

SUMMARY OF CHAPTER 2

Positive protons and neutral neutrons are held together in the nucleus of an atom.

The nucleus is therefore positively charged. It holds nearly all the mass of an atom.

Electrons orbit in the space around the nucleus, but only with certain specific energies known as energy levels.

Each energy level can hold a definite maximum number of electrons.

The number of protons in the nucleus is the atomic number. This is equal to the number of electrons orbiting the nucleus.

Each element has a distinct atomic number.

The Periodic Table is an arrangement of elements in order of increasing atomic number.

Elements in the same vertical group of the Table have similar properties, and their atoms have the same number of outer electrons.

Many properties of an atom are governed by the number of outer electrons.

Atoms with eight outer electrons (or two for the first energy level) are stable.

CHAPTER 3

CHEMICAL BONDING

Petrol burning, a nail rusting, or a cake baking are all *chemical reactions.* Chemical reactions usually change substances dramatically. For example when iron rusts:

| *Before rusting* Shiny, strong iron in moist air | chemical ⟶ reaction | *After rusting* Crumbly brown powder called rust |

The iron and the constituents of moist air are still in rust, but not just mixed together. They are bound strongly together by *chemical bonds.* A chemical reaction is no more than a rearrangement of the chemical bonds which hold the various atoms together. To understand chemical reactions, we must picture how elements bond.

3.1 COMPOUNDS

Different elements chemically *bonded* together rather than simply *mixed* together are called *compounds.*

Water is a compound of the elements hydrogen and oxygen, carbon dioxide is a compound of the elements carbon and oxygen, and rust (hydrated iron oxide) is a compound of iron, oxygen and hydrogen.

A compound has very different properties from its constituent elements. The compound water is quite different from a mixture of hydrogen gas and oxygen gas because, in water, hydrogen and oxygen atoms are *chemically bonded* together. Similarly a fine mixture of powdered carbon with oxygen gas bears no resemblance to the compound carbon dioxide, in which carbon and oxygen are *chemically bonded.*

Why should atoms form chemical bonds? It is suggested that the answer is *stability.*

3.2. STABILITY

Many things that 'just happen'—an apple falling from a tree, water flowing to the sea, a clock spring steadily unwinding—involve a loss of potential energy. A slate falls from a roof on to the ground (a position of lower potential energy) with a bang. Some potential energy has been given up as sound energy. The slate is now in a *more stable* position.

A pen balanced on its point will fall over on to its side to a *more stable* position of lower potential energy. It seems that happenings are likely to happen if potential energy is lost, with the objects involved becoming more stable.

Section 2.7 showed that, for atoms, stability means having the electron arrangement of an inert gas. Any atom could become more stable by

giving electrons to,

taking electrons from, or

sharing electrons with

another atom in order to achieve the electron structure of an inert gas.

Bonding is the result of an atom's *quest for stability*. The following sections outline theories or pictures to explain bonding. Their effectiveness may then be tested against the chemical behaviour outlined later.

3.3 IONIC BONDING

(a) Sodium chloride

Sodium is an extremely reactive, dangerous metal. Large lumps of it will explode on contact with water. Chlorine is a choking green gas once used as a wartime poison. If sodium is dropped into chlorine gas, a violent reaction ensues. However, the resulting white solid compound of these two elements is common table salt, which is not only harmless but essential to our survival. Why the change, and why the reaction?

Atoms of sodium and chlorine have the following electron arrangements

$$_{11}Na \quad 11p \ 2e \ 8e \ 1e$$
$$_{17}Cl \quad 17p \ 2e \ 8e \ 7e$$

Neither atom is stable. By losing its one outer electron the sodium atom could find stability, for it would then be like the inert gas neon, with an electron arrangement 2e 8e. By taking in the lost electron the chlorine atom could similarly achieve stability with the electron arrangement of argon, 2e 8e 8e. By losing and gaining electrons, the atoms now find themselves with a net charge. They have become *ions*.

Atoms   Ions

$_{11}$Na   11p 2e 8e(1e) ⟶ $_{11}$Na   11p 2e 8e

$_{17}$Cl   17p 2e 8e 7e ⟶ $_{17}$Cl   17p 2e 8e 8e

The sodium ion still has 11 protons (+) but now only 10 electrons (−). There is a 'left-over' 1+ charge, and the ion is written as Na^+.

The chlorine ion still has 17 protons (+) but now 18 electrons (−). There is a 'left-over' 1− charge, and the ion is written as Cl^-.

Through the movement of electrons, both sodium and chlorine have become stable *ions*. Their opposite charges bind them together (opposite charges attract). An *ionic bond* is no more than this plus-to-minus attraction. In a crystal of sodium chloride, this ionic bonding continues from one oppositely charged ion to the next throughout the solid structure (Fig. 3.1).

fig 3.1 *continuous ionic bonding in sodium chloride*

Na^+	Cl^-	Na^+	Cl^-	Na^+	Cl^-	Na^+
Cl^-	Na^+	Cl^-	Na^+	Cl^-	Na^+	Cl^-
Na^+	Cl^-	Na^+	Cl^-	Na^+	Cl^-	Na^+
Cl^-	Na^+	Cl^-	Na^+	Cl^-	Na^+	Cl^-
Na^+	Cl^-	Na^+	Cl^-	Na^+	Cl^-	Na^+
Cl^-	Na^+	Cl^-	Na^+	Cl^-	Na^+	Cl^-
Na^+	Cl^-	Na^+	Cl^-	Na^+	Cl^-	Na^+

Sodium *ions* and chloride *ions* have quite different properties from *atoms* of sodium and chlorine. As ions they hold eight electrons in their outer energy levels, and are now as unwilling to react as neon and argon respectively. Eating sodium chloride is therefore closer to swallowing a mixture of two inert gases than to gulping down a corrosive plateful of sodium and chlorine.

(b) Magnesium oxide
Magnesium metal burns in air with a brilliant flame, reacting with oxygen to form ionically bonded magnesium oxide as a white powder.

Atoms   Ions

$_{12}$Mg   12p 2e 8e(2e) ⟶ $_{12}$Mg   12p 2e 8e

$_8$O   8p 2e 6e ⟶ $_8$O   8p 2e 8e

The magnesium ion still has 12 protons (+) but now only 10 electrons (−). There is a 'left-over' 2+ charge, and the ion is written as Mg^{2+}.

The oxide ion still has 8 protons (+) but now 10 electrons (−). There is a 'left-over' 2− charge, and the ion is written as O^{2-}.

Thus magnesium oxide consists of oppositely charged Mg^{2+} and O^{2-} ions ionically bonded around one another.

(c) Lithium oxide

When a lump of the metal lithium is heated in air, it suddenly bursts into a deep crimson flame, reacting with oxygen to form the white ionic solid lithium oxide.

By losing one electron, a lithium atom can achieve the stable electron arrangement of helium (the first energy level is stable with just two electrons).

$$Atom \qquad\qquad Ion$$
$$_3Li \quad 3p\ 2e\ 1e \longrightarrow\ _3Li \quad 3p\ 2e \ (or\ Li^+)$$

However, oxygen needs to gain *two* electrons to achieve stability.

$$Atom \qquad\qquad Ion$$
$$_8O \quad 8p\ 2e\ 6e \longrightarrow\ _8O \quad 8p\ 2e\ 8e\ (or\ O^{2-})$$

There is no evidence of unwanted electrons drifting aimlessly around looking for a reaction. How, then, can oxygen atoms find two electrons when each lithium atom gives up only one? The answer is that for every oxygen atom requiring two electrons, *two* lithium atoms react, each giving away one electron.

$$Atoms \qquad\qquad Ions$$
$$_3Li \quad 3p\ 2e(1e) \qquad\qquad _3Li \quad 3p\ 2e \quad (Li^+)$$
$$_3Li \quad 3p\ 2e(1e) \longrightarrow\ _3Li \quad 3p\ 2e \quad (Li^+)$$
$$_8O \quad 8p\ 2e\ 6e \qquad\qquad _8O \quad 8p\ 2e\ 8e \quad (O^{2-})$$

Thus solid lithium oxide contains two lithium ions for every one oxygen ion, written as $Li^+_2\ O^{2-}$ or more simply as Li_2O. When the formula of lithium oxide is experimentally determined, this same answer is obtained. This lends further support to the theory of atomic structure of Chapter 2. Not only does it provide us with an explanation of how lithium and oxygen react, but it also enables us to predict a formula for lithium oxide which agrees with experiment (see later).

(d) Calcium fluoride

The metal calcium bursts into flame in fluorine gas to form calcium fluoride. This ionic solid is the 'fluoride' added to many toothpastes in order to strengthen tooth enamel. Two electrons are lost as each calcium atom forms a stable Ca^{2+} ion, whilst a fluorine atom requires only one electron to become stable. Each calcium atom therefore reacts with two fluorine atoms.

$$
\begin{array}{ll}
\textit{Atoms} & \textit{Ions} \\
{}_{20}\text{Ca} \quad 20p \ 2e \ 8e \ 8e\{2e\} & {}_{20}\text{Ca} \quad 20p \ 2e \ 8e \ 8e \\
{}_{9}\text{F} \quad 9p \ 2e \ 7e & {}_{9}\text{F} \quad 9p \ 2e \ 8e \\
{}_{9}\text{F} \quad 9p \ 2e \ 7e & {}_{9}\text{F} \quad 9p \ 2e \ 8e
\end{array}
$$

Hence calcium fluoride has the formula $Ca^{2+} F^-_2$ or more simply CaF_2.

(e) Other examples

Using Table 2.2 of electronic structures, the reader should now be able to explain the reactions and verify the formulae when

(i) sodium reacts with sulphur to form sodium sulphide, Na_2S;

(ii) magnesium reacts with chlorine to form magnesium chloride, $MgCl_2$;

(iii) potassium reacts with nitrogen to form potassium nitride, K_3N.

(f) Ionic bonding and the Periodic Table

Elements forming positive ions have few outer electrons to lose. Such elements are found to the left of the Periodic Table and are classed as metals. Negative ions are derived from atoms which already have close to eight outer electrons. These are the non-metals, and they are found to the right of the Table.

Therefore ionic compounds are formed when an element from the left of the Periodic Table reacts with one from the right. Furthermore, as a rough rule, the more extreme the separation of the elements in the Table, the more vigorous the ion-forming reaction between them.

3.4 COVALENT BONDING

A quick glance at the Periodic Table (Table 2.3) will show that sulphur and oxygen, phosphorus and chlorine, and chlorine and fluorine are very closely positioned, and that they are all on the right-hand side of the Table. It is difficult to see how any of these elements could be persuaded to lose electrons. *Both* members of each pair achieve stability only through electron *gain*. Since neither atom will lose electrons, neither atom is in a position to gain electrons. Nevertheless these pairs of elements do react vigorously together to form strongly bonded compounds. In these examples, however, there is no evidence of electrons jumping from one atom to another to form ions: here the bonding is pictured as being *covalent*.

(a) Hydrogen chloride

An explosion occurs as soon as a mixture of hydrogen and chlorine is exposed to ultra-violet light. The explosion product is hydrogen chloride gas.

Hydrogen atoms with the structure ${}_1\text{H}$ $1p$ $1e$ each need to *gain* one electron to attain the stable electron arrangement of helium (the first energy level is filled and stable with just two electrons). Similarly each chlorine atom $_{17}$Cl 17p 2e 8e 7e needs to *gain* one electron to achieve stability. Since both atoms need electrons, both could become stable if each *shared* an electron with the other.

By sharing an electron from the chlorine atom, hydrogen can achieve the stability of two electrons (one of its own and one shared) in its first energy level. By sharing the one electron of hydrogen, chlorine effectively possesses eight electrons (seven of its own and one shared) in its outer energy level.

Covalent bonding is the result of electron sharing. It is helpful to represent this sharing diagrammatically. Dots, crosses or open circles are used to represent the electrons only of the outermost energy level.

Chlorine $_{17}$Cl 17p 2e 8e 7e is drawn

$$\begin{array}{c} \text{x x} \\ \text{x Cl x} \\ \text{x x} \end{array}$$

Hydrogen $_1$H 1p 1e is drawn

$$\text{H} \bullet$$

The sharing of electrons is then shown as

$$\begin{array}{c} \text{x x} \\ \text{H} \overset{\text{x}}{\bullet} \text{Cl} \overset{\text{x}}{} \\ \text{x x} \end{array}$$

Only the two shared electrons will be drawn between the atoms.

Pairs of dots or crosses *between* atoms always denote *shared electrons*. The pair of shared electrons between the hydrogen and the chlorine atom comprise *one covalent bond*. Both atoms are held together in order that, by jointly sharing electrons, they may each achieve a stable electron arrangement.

A covalent bond may also be represented by a single line between the atoms

One covalent bond/Shared pair of electrons

(b) Chlorine

Experiment shows that chlorine gas does not consist of individual chlorine atoms, but rather of pairs of atoms. A single chlorine atom holds seven outer electrons. Through covalently sharing one pair of electrons, two atoms can achieve stability, since each now effectively holds eight outer electrons (seven of its own and one shared).

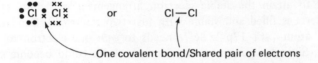

One covalent bond/Shared pair of electrons

(c) Molecules

A group of covalently bound atoms is a *molecule*. Both hydrogen chloride, HCl, and chlorine gas, Cl_2, are examples of *diatomic molecules*. Water, H_2O, is a *triatomic molecule*, whilst hydrogen peroxide, H_2O_2, with four atoms in each molecule, is *tetratomic*. The gas helium, He, is *monatomic*.

(d) Electron pairs

In the electron dot and cross diagrams above, the electrons are deliberately drawn in pairs. We have already seen that a covalent bond is a *pair* of shared electrons, and that these are drawn between the atoms concerned. The non-bonding electrons, those which are not involved in sharing, are similarly drawn in pairs called *lone-pair electrons*. There are therefore shared pairs of electrons between atoms, and lone-pair electrons elsewhere around the atom.

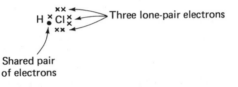

Three lone-pair electrons

Shared pair
of electrons

The theoretical reason for electron pairing is complex and beyond the scope of this book. A more practical justification is that, by assuming that electrons *do* pair, molecular shapes can be predicted which are in good agreement with experiment.

(e) Further examples

(i) *Water*

The formula of water is H_2O. Oxygen, $_8O$ 8p 2e 6e, needs to share two more electrons; hydrogen, $_1H$ 1p 1e, requires only one.

$$H \overset{\bullet\bullet}{\underset{\bullet\bullet}{\overset{\times}{O}}} H \qquad \text{or} \qquad H - \overset{\bullet\bullet}{\underset{\bullet\bullet}{O}} - H$$

In water, the oxygen atom holds two lone-pair electrons.

(ii) *Methane*

The formula of natural gas, methane, has been determined as CH_4. Carbon,

$_6$C 6p 2e 4c, needs to share four electrons to become stable. Hydrogen, $_1$H 1p 1e, needs to share one electron. Hence we can write

Methane has no lone-pair electrons.

(iii) *Ammonia*

The molecular formula of ammonia, an important ingredient of fertilisers, has been determined as NH_3. Nitrogen, $_7$N 7p 2e 5e, needs to share three electrons. Hydrogen, $_1$H 1p 1e, needs to share one electron. Hence we can write

Each ammonia molecule has one lone pair of electrons.

(f) Multiple bonding

Sometimes several pairs of electrons are shared between two atoms, giving rise to double or triple covalent bonds. These are illustrated in the following examples.

(i) *Oxygen*

Oxygen gas has been shown to exist as diatomic O_2 molecules. $_8$O 8p 2e 6e, each oxygen atom needs to share two electrons. Here the oxygen atoms each share *two* pairs of electrons to achieve stability.

$$\overset{\bullet\bullet}{\underset{\bullet\bullet}{O}}\overset{\bullet}{\underset{\bullet}{\vdots}}\overset{\times}{\underset{\times}{O}}\overset{\times}{\underset{\times}{}}$$

Only the shared electrons will be drawn between the two atoms. By this double sharing, each atom effectively holds eight outer electrons (six of its own, and two shared). Using the line representation, the oxygen molecule is drawn as

$$\overset{\bullet\bullet}{\underset{\bullet\bullet}{O}}=\overset{\times}{\underset{\times}{O}}\overset{\times}{\underset{\times}{}}$$

The atoms are held by a *double covalent bond.* Double bonds are stronger than single bonds. Doubly bonded atoms are therefore pulled closer to one another.

(ii) *Carbon dioxide*

Carbon dioxide molecules, with the formula CO_2, contain two double bonds.

$$\overset{\times}{\underset{\times}{\overset{\times}{O}}}\overset{\times}{\underset{\times}{\times}}\;\overset{\bullet}{\underset{\bullet}{C}}\;\overset{\times}{\underset{\times}{\times}}\overset{\times}{\underset{\times}{\overset{\times}{O}}} \qquad \text{or} \qquad \overset{\times\;\times}{\underset{\times\;\times}{O}}{=}C{=}\overset{\times}{\underset{\times}{\overset{\times}{O}}}$$

(iii) *Nitrogen*

Diatomic nitrogen molecules provide an example of triple covalent bonding.

$$\overset{\bullet}{\underset{\bullet}{}}N\;\overset{\bullet}{\underset{\bullet}{}}\;\overset{\times}{\underset{\times}{}}N\;\overset{\times}{\underset{\times}{}} \qquad \text{or} \qquad \overset{\bullet}{\underset{\bullet}{}}N\equiv N\;\overset{\times}{\underset{\times}{}}$$

(g) Further examples

Using Table 2.2 to establish the number of outer electrons, the reader should now be able to draw electron dot and cross diagrams showing how the stable inert-gas configuration is attained in the following covalent molecules:

  (i) nitrogen trifluoride, NF_3;
  (ii) tetrachloromethane (carbon tetrachloride), CCl_4;
  (iii) hydrogen sulphide, H_2S;
  (iv) carbon disulphide, CS_2.

3.5 SHAPES OF MOLECULES

Although far too small to be seen, the shapes of many molecules have been determined from measurements of the way in which X-rays are affected when travelling through them. This technique of 'looking at' atoms, called *X-ray diffraction spectroscopy*, involves complex mathematical calculations which are usually carried out by computer.

However, we can make accurate guesses at molecular shapes using the picture of covalent bonding outlined in Section 3.4, simply by assuming that groups of electrons around an atom will be mutually repelled as far from each other as possible. Any covalent bond between two atoms, whether single, double or triple, is counted as one group of electrons. Lone-pair electrons also count as one group.

For example, carbon disulphide, CS_2, is represented as

$$\overset{x}{\underset{x}{x}}S\overset{x}{\underset{x}{x}}\bullet\overset{\bullet}{\bullet}C\overset{\bullet}{\bullet}\overset{x}{\underset{x}{x}}S\overset{x}{\underset{x}{x}}$$

Here the carbon atom is surrounded by two groups of electrons (two double bonds). Arranging two groups of electrons as far from each other as possible would give a linear pattern as shown in Fig. 3.2. The thick lines of the diagram represent electron groups. Carbon disulphide and carbon dioxide are both linear molecules.

fig 3.2   *the linear structure giving most separation of two repelling*
          groups

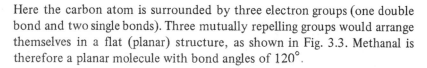

Methanal, H_2CO, is used in aqueous solution to preserve biological specimens. Often still called 'formalin' or 'formaldehyde' in biological circles, its structure is

Here the carbon atom is surrounded by three electron groups (one double bond and two single bonds). Three mutually repelling groups would arrange themselves in a flat (planar) structure, as shown in Fig. 3.3. Methanal is therefore a planar molecule with bond angles of $120°$.

fig 3.3   *the planar structure giving most separation of three*
          mutually repelling groups

In the ammonia molecule, the atom of nitrogen has four groups of electrons around it

$$H \overset{x}{\underset{\bullet}{\bullet}} \overset{\bullet\bullet}{N} \overset{\bullet}{\underset{x}{x}} H$$
$$H$$

(three single bonds and one lone pair of electrons).

Similarly the central atoms of the water molecule and the methane molecule are surrounded by four electron groups:

$$H \overset{x}{\underset{\bullet}{\bullet}} \overset{\bullet\bullet}{\underset{\bullet\bullet}{O}} \overset{\bullet}{\underset{x}{x}} H$$

$$H$$
$$H \overset{x}{\underset{x}{\bullet}} \overset{\bullet x}{\underset{x\bullet}{C}} \overset{\bullet}{\underset{x}{x}} H$$
$$H$$

Water   Methane

The spacial arrangement around these atoms is the most difficult to visualise. If the structure were flat, then the planar shape illustrated below would be found.

By allowing the electron groups to bend away from a planar arrangement, the structure illustrated in Fig. 3.4 gives the greatest group separation. Here the angles between groups are all 109.4°. The shape is called

fig 3.4 *the tetrahedral structure giving most separation of four mutually repelling groups*

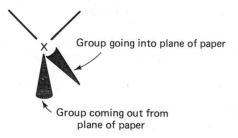

Group going into plane of paper

Group coming out from plane of paper

tetrahedral, and the angle of 109.4° the *tetrahedral angle*. The word 'tetrahedral' reflects the fact that joining the outer corners produces a four-sided figure, or tetrahedron.

Methane is therefore a tetrahedral molecule

with bond angles of 109.4°.

Although lone-pair electrons must be included when *predicting* molecular shape, when *describing* the shape of a molecule only the atoms are considered. Thus water is described as having bent planar molecules

even though lone-pair electrons come out from and go into the plane. Similarly, describing the shape of ammonia, NH_3, but ignoring the lone pair of electrons, leaves us with a *pyramid* with bond angles of 109.4°.

or

The reader should now be able to determine the shape of:
(i) hydrogen sulphide, H_2S, as bent planar;
(ii) tetrafluoromethane, CF_4, as tetrahedral;
(iii) thiomethanal, H_2CS, as planar with bond angles of 120°.

3.6 METALLIC BONDING

Atoms situated in opposite halves of the Periodic Table bond ionically and atoms both situated to the right of the Table bond covalently. How, then, do atoms from the left of the Periodic Table, the metals, bond with one another? What sort of bonding is there between the atoms in a solid lump of metal? A strong bond certainly is formed: aluminium aeroplane wings, ancient bronze spearheads, and the steel girders of bridges and office blocks depend for their strength and toughness upon the bond that forms between metal atoms.

This is *metallic bonding*. Stability is not attained in this case through the inert-gas structure, but by the phenomenon of electron spread or *delocalisation*.

Imagine a small box filled with twenty-million electrons. Since the electrons are all negatively charged, they mutually repel. Upon opening the lid, electrons would be repelled out at high speed (Fig. 3.5). The electrons were in a position of high potential energy and relatively unstable. If these twenty-million electrons were now put into a larger box, they would be less unstable. Being able to get further away from one another, they would experience less repulsion, and would fly out with less energy were the lid to be undone (Fig. 3.6).

32

fig **3.5** *release of electrons from a small box*

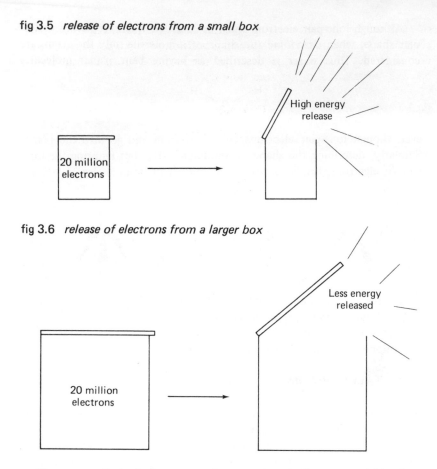

fig **3.6** *release of electrons from a larger box*

The more that electrons can be spread out, the more stable they become. This 'spreading out' of electrons is called *delocalisation*, and the resulting stability is called *delocalisation stability*.

Metals are elements which can lose their outer electrons. For example, sodium will easily lose its one outer electron to form the Na^+ ion.

$$\text{Atom} \qquad\qquad\qquad \text{Ion}$$

$$_{11}Na \;\; 11p \; 2e \; 8e \; 1e \quad \xrightarrow[\text{loses 1 e}]{\text{easily}} \quad _{11}Na \;\; 11p \; 2e \; 8e$$

Aluminium atoms readily lose three electrons each.

$$\text{Atom} \qquad\qquad\qquad \text{Ion}$$

$$_{13}Al \;\; 13p \; 2e \; 8e \; 3e \quad \xrightarrow[\text{loses 3 e}]{\text{easily}} \quad _{13}Al \;\; 13p \; 2e \; 8e$$

If many metal atoms are brought close together, the outer energy levels of each can be pictured merging together as these levels begin to overlap (Fig. 3.7). Outer electrons are then in a position to move not just around one atom but around and between all the atoms. These electrons have become *delocalised* and therefore *more stable*.

fig 3.7 *the formation of metallic bonds*

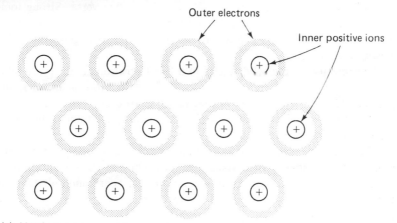

(a) Metal atoms approaching

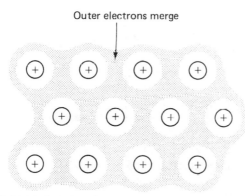

(b) Metal atoms bonding

This is *metallic bonding* in which, by being together, the atoms create a more stable arrangement. A block of metal should now be visualised as millions of metal ions, like tiny islands, in a sea of delocalised electrons. The free movement of electrons through and around a metal offers a good explanation for the ease with which all metals conduct electricity.

Ionic, covalent and metallic bonds comprise the strongest bonds within chemicals. They are called *primary bonds*.

3.7 GIANT STRUCTURES

(a) Giant ionic structures

We have seen that a lump of sodium chloride consists of billions of Na^+ and Cl^- ions (Section 3.3(a)). There is no such thing as one individual molecule of sodium chloride in this solid lump, because any one Na^+ ion will attract around it many Cl^- ions, and vice versa. Strong ionic bonding continues throughout the solid crystal from one ion to the next. When primary bonding is continuous in this way, the arrangement is called a *giant structure* (Fig. 3.8).

Any one ion in the solid is held in a fixed position to all surrounding ions by strong ionic bonds. The ions are fixed in position in a definite pattern which repeats regularly throughout the structure. Ionic compounds are therefore *crystalline solids*. Looking at any ionic solid under the microscope reveals its regular crystalline nature.

When an ionic solid is heated, we imagine the heat energy to vibrate the ions more and more in their fixed positions. Only at quite high temperatures is there sufficient energy to vibrate the ions apart, and only then are the ions free to move around and between one another as a liquid. Thus ionic solids have high melting points (see Table 3.1).

fig 3.8 *arrangement of ions in a giant structure*

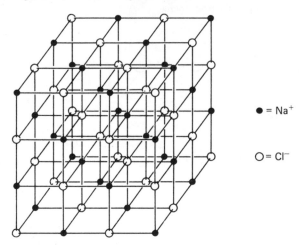

$\bullet = Na^+$

$O = Cl^-$

(b) Giant metallic structures

With the exception of mercury, which is a dense molten metal even at room temperature, metals are solids with generally high melting points (see Table 3.1). Like ionic compounds, a high melting point suggests strong

Table 3.1   *the melting point and structure of various substances*

Substance	Melting point/°C	Type of structure
Sodium chloride	808	giant ionic
Calcium oxide	2600	giant ionic
Iron	1539	giant metallic
Calcium	850	giant metallic
Potassium	63	giant metallic
Copper	1083	giant metallic
Sulphur	114	molecular
Oxygen	− 219	molecular
Bromine	− 7	molecular
Diamond	3550	giant covalent

continuous bonding from one atom to the next throughout the solid, and metals too have *giant structures*.

Strength of metallic bonding is related to the size of metal atoms. Small atoms such as iron or copper pack together closely to form strongly bonded structures with very high melting points. In contrast, the melting points of sodium and potassium, with relatively large atoms, are comparatively low. Metallic bond strength is also related to the number of delocalised outer electrons. Calcium ($_{20}$Ca 20p 2e 8e 8e 2e), with two delocalised electrons per atom, melts at a higher temperature than potassium ($_{19}$K 19p 2e 8e 8e 1e) which has only one.

(c) Giant covalent structures

Ionic and metallic bonding always creates giant structures. Covalently bonded atoms usually form into molecular structures (see Section 3.8). Nevertheless, a few substances are known in which the atoms are held in a continuous giant structure by covalent bonds, forming solids with exceptionally high melting points. Diamond and graphite are examples of such *giant covalent structures*. Both are differently bonded arrangements of covalently bonded carbon atoms (Fig. 3.9).

Carbon ($_{6}$C 6p 2e 4e), with four electrons involved in bonding, should form four bonds per atom. This is seen to be the case in diamond, but in graphite each carbon atom is shown with only three bonds. Since graphite conducts electricity with ease (it is the only non-metal to do this), it is believed that the fourth electrons are spread out or *delocalised* over the entire giant structure.

fig 3.9 *covalently bonded giant structures of (a) diamond and (b) graphite*

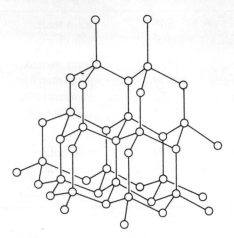

O = Carbon atom

(a) Diamond

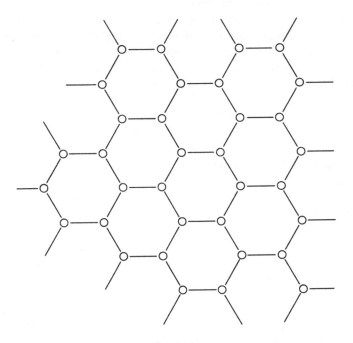

O = Carbon atom

(b) Graphite

3.8 MOLECULAR STRUCTURES

(a) Secondary bonding

Ice is a solid made up of covalently bonded H_2O molecules. Yet, even at room temperature, bonds have apparently broken to leave liquid water. Covalently bonded molecules of CO_2 form a solid at $-80°C$, but at room temperature some bonds must have broken since we find carbon dioxide to be a gas. In both examples, bonds *have* broken as the solids melt or vaporise. These broken bonds cannot be covalent; covalent bonds could be split only by considerably greater energies.

Within molecules the individual atoms are joined by strong covalent bonds, but *between* neighbouring molecules only exceptionally weak forces of attraction exist. They are called *secondary bonds* (Fig. 3.10). They are usually at least one-hundred times weaker than the strong covalent bonds.

fig 3.10 *secondary bonding between molecules*

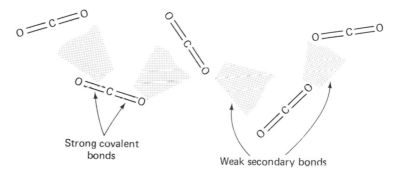

Secondary bonds are a consequence of weak electrical attractions between charged particles within atoms of neighbouring molecules.

Covalent solids are generally held together in a *molecular structure* by these weak secondary bonds, and very little energy is needed to overcome this attraction.

(b) Hydrogen bonding

The strongest secondary bonding is found to occur between molecules containing covalent

    nitrogen-hydrogen bonds,

    oxygen-hydrogen bonds,

    chlorine-hydrogen bonds and

    sulphur-hydrogen bonds.

For example, the attraction between water molecules is greater than might be expected. Water is a liquid at ordinary temperatures as a result of

38

particularly strong secondary bonds between its molecules. Because of the involvement of hydrogen it is described as *hydrogen bonding*.

This hydrogen bonding is believed to result from a greater pull of electrons towards nitrogen, oxygen, chlorine or sulphur atoms than towards atoms of hydrogen (the atoms are said to be more *electronegative* than hydrogen, see Section 10.1(b)). A slight negative charge (represented by $\delta-$) is therefore gained by these atoms, leaving hydrogen slightly positive ($\delta+$). Thus water has a slight charge separation; it is a *polar* molecule (Fig. 3.11). Attractions between such molecules form hydrogen bonds (Fig. 3.12).

fig 3.11 *separation of charge in the water molecule*

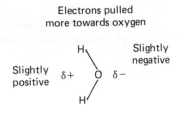

Electrons pulled
more towards oxygen

Slightly
negative

Slightly
positive

fig 3.12 *secondary hydrogen bonding between water molecules*

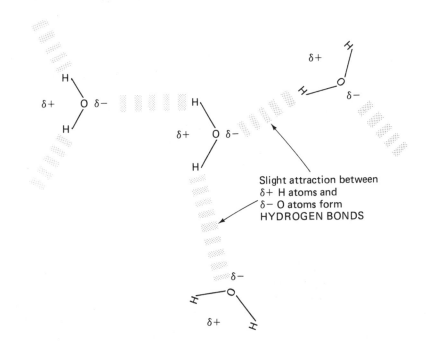

Slight attraction between
$\delta+$ H atoms and
$\delta-$ O atoms form
HYDROGEN BONDS

Although the strongest of the secondary bonds, hydrogen bonding is nevertheless in the region of twenty times weaker than most primary bonds.

Being held only by secondary bonds, molecular structures melt at low temperatures. They are often liquids or gases even at ordinary temperatures. In Table 3.1 the molecular structures can be identified easily by their lower melting points.

3.9 ELECTRICAL CONDUCTION

Electricity flows when charged particles move. Our picture of metals (and graphite) with delocalised freely moving electrons is consistent with their ability to conduct electricity.

Ionic compounds contain charged particles—positive and negative ions—but these are rigidly held in a giant array. Ionic solids are poor conductors because the ions are unable to move freely. Melting an ionic compound, or dissolving it in water, gives its ions the freedom to move, and then electricity will flow through (see Section 8.7).

Table 3.2 *dependence of properties upon structure*

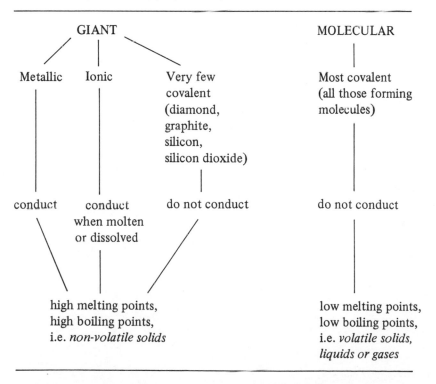

Covalent compounds, whether solid, liquid or gas, have no net charge; they are all non-conductors.

SUMMARY OF CHAPTER 3

Atoms bond together as a way of achieving stability.

If an atom A bonds with an atom X, stability is found in one of three ways:

(i) *If A is to the left of the Periodic Table and X to the right.* A loses electrons to X in order that both might achieve an inert-gas electron structure. The resulting A^+ and X^- ions are electrically attracted. This is *ionic bonding.*

(ii) *If A and X are both to the right of the Periodic Table.* Outer electrons of A and X share in order that both might achieve an inert-gas electron structure. This is *covalent bonding.*

(iii) *If A and X are both to the left of the Periodic Table.* The atoms are pulled together in a *metallic bond* in which the outer electrons gain stability through delocalisation.

Some properties depend upon structure, as shown in Table 3.2.

CHAPTER 4

THE BASIC LANGUAGE AND TECHNIQUES OF CHEMISTRY

4.1 CHEMICAL VOCABULARY

Names, formulae, equations and quantities comprise the language of chemistry.

It should be realised that by using rules and theories we can at best only *predict* formulae. If the predictions agree with experiment then the rules and theories are acceptable, but formulae can be truly determined only by *carrying out experiments*.

(a) Ionic compounds
Section 3.3 showed how the formulae of an ionic compound could be predicted by

(i) writing down the electronic structure of the ions concerned,
(ii) deciding how many electrons would be lost and gained,
(iii) adjusting the number of atoms so that number of electrons lost = number of electrons gained.

This laborious process can be avoided once the reader has become familiar with charges on common ions (Tables 4.1 and 4.2).

A routine for quickly predicting an ionic formula is given below.

(i) *Sodium nitride*

sodium ion charge is +1   nitride ion charge is -3

hence 3 sodium ions will combine with every 1 nitride ion
formula: NaN_3.

Table 4.1 *charges and names* of common positive ions*

Charge	1+	2+	3+
	Li^+ lithium	nearly all other metals	
	Na^+ sodium		Al^{3+} aluminium
	K^+ potassium		
	Ag^+ silver	Fe^{2+} iron(II)	Fe^{3+} iron(III)
	Cu^+ copper(I)	Cu^{2+} copper(II)	
	NH_4^+ ammonium		

*Note on names given in table. Roman numerals in the name indicate the charge on the ion. Thus sodium chloride should be written sodium(I) chloride and aluminium oxide as aluminium(III) oxide. In practice if the ionic charge is obvious roman numerals are often omitted. However, when atoms can form differently charged ions it is essential to include the numerals in order to distinguish between, for example, iron(II) oxide containing Fe^{2+} ions and iron(III) oxide containing Fe^{3+} ions.

(ii) *Aluminium sulphide*

aluminium ion charge is +3   sulphide ion charge is 2

hence 2 aluminium ions will combine with every 3 sulphide ions
formula: Al_2S_3.

(iii) *Iron(III) sulphate*

iron(III) ion charge is +3   sulphate ion charge is −2

hence 2 iron(III) ions will combine with every 3 sulphate ions
formula: $Fe_2(SO_4)_3$.

Table 4.2 *charges and names* of common negative ions*

Charge	3–	2–	1–
	N^{3-} nitride	O^{2-} oxide	F^- fluoride
		S^{2-} sulphide	Cl^- chloride
			Br^- bromide
			I^- iodide
	PO_4^{3-} phosphate(V)	SO_3^{2-} sulphite	OH^- hydroxide
		SO_4^{2-} sulphate	NO_2^- nitrite
			NO_3^- nitrate
			MnO_4^- manganate(VII) (or permanganate)

*Note on names given in table. The ending '-ate' in the name of an ion signifies that oxygen is bound within the ion. The ending '-ite' indicates less oxygen. In advanced work, sulphate should strictly be tetraoxo-sulphate(VI), nitrate should be trioxonitrate(V), etc. The naming system used here is that suggested for non-advanced work by the Association for Science Education in their report *Chemical Nomenclature, Symbols and Terminology.*

(iv) *Copper(II) chloride*

copper(II) ion charge is +2   chloride ion charge is -1

hence 1 copper(II) ion will combine with every 2 chloride ions
formula: $CuCl_2$.

(b) Covalent compounds

Once the reader has become familiar with the number of electrons generally shared by common non-metals (see Table 4.3), formulae of covalent compounds can be predicted by a routine similar to the one described above for ionic compounds.

Table 4.3 *number of electrons commonly shared by non-metals*

Number of electrons shared	4	3	2	1
	C	N	O	F
	Si	P*	S*	Cl
				I
Periodic Table group	4	5	6	7

*These elements commonly share more electrons, since their outer (third) energy level can hold up to 18 electrons. Thus phosphorus shares 3, but can also share 5 electrons; and sulphur shares 2, but can also share 4 or 6 electrons.

The name describes the molecular composition. Thus PCl_3 is phosphorus trichloride and PCl_5 is phosphorus pentachloride. Sometimes it is impossible to be precise about the structure of a molecule. One oxide of phosphorus, for example, has molecules with formulae somewhere in between P_2O_3 and P_4O_6. In such a case, roman numerals are used to denote the number of electrons shared by the central atom. This oxide is phosphorus(III) oxide. Similarly the oxide with molecules somewhere in between P_2O_5 and P_4O_{10} is phosphorus(V) oxide.

Table 4.3 shows a clear relationship between the number of electrons shared and Periodic Table group number.

The routine for quickly predicting a covalent formula is given below.

(i) *Methane*

carbon atoms share 4 electrons   hydrogen atoms share 1 electron

hence 1 atom of carbon will share with 4 atoms of hydrogen
formula: CH_4.

(ii) *Water*

oxygen atoms share 2 electrons   hydrogen atoms share 1 electron

hence 1 atom of phosphorus will share with 3 atoms of chlorine
formula: PCl_3.

(iii) *Phosphorus trichloride*

phosphorus atoms share 3 electrons   chlorine atoms share 1 electron

hence 1 atom of phosphorus will share with 3 atoms of chlorine
formula: PCl_3.

(iv) *Phosphorus pentachloride*

phosphorus atoms share 5 electrons   chlorine atoms share 1 electron

hence 1 atom of phosphorus will share with 5 atoms of chlorine
formula: PCl_5.

(c) Writing equations

When powdered aluminium is heated in a strong test tube with powdered
sulphur, aluminium sulphide is produced in a violent eruption. We can
write this as

aluminium + sulphur $\longrightarrow$ aluminium sulphide

Symbols could be used to replace the words in this chemical *equation*.
Aluminium is $Al(s)$, where (s) indicates solid (similarly (l) is liquid, (g)
gas, (aq) aqueous solution, (alc) alcoholic solution). Sulphur is $S(s)$ and
aluminium sulphide is $Al_2 S_3 (s)$.

Using these symbols, we might write

$$Al(s) + S(s) \longrightarrow Al_2 S_3 (s)$$

However the formula of aluminium sulphide indicates that two aluminium
and three sulphur atoms are required. The equation must be made to
balance, with the same number of the same atoms before and after reaction, so we must have

$$2Al(s) + 3S(s) \longrightarrow Al_2 S_3 (s)$$

Note that when balancing an equation only the numbers *in front of*
the chemical symbols can be altered, the chemical formulae themselves
remain unchanged.

When nitrogen reacts with hydrogen, the gas ammonia is obtained.
Just writing down the formulae gives

$$N_2 (g) + H_2 (g) \longrightarrow NH_3 (g)$$

Since N_2 molecules contain two nitrogen atoms, two molecules of
ammonia must be produced, $2NH_3$. These contain six atoms of hydrogen,
indicating the need for six hydrogen atoms, that is $3H_2$, at the start.
Hence we have

$$1N_2 (g) + 3H_2 (g) \longrightarrow 2NH_3 (g)$$

A similar process could now be used to balance the following equations.

(i) $Cu(s) + O_2 (g) \longrightarrow CuO(s)$
(ii) $Al(s) + Cl_2 (g) \longrightarrow AlCl_3 (s)$
(iii) $SO_2 (g) + H_2 S(g) \longrightarrow H_2 O(l) + S(s)$

(d) The mole

In reality we are concerned not with individual particles but with several million million million particles. For example, one gramme of water contains about thirty thousand million million million molecules. To measure out one molecule of water is impossible, but to measure out 600 000 000 000 000 000 000 000 or 6×10^{23} molecules is easy, since they happen to weigh 18 grammes. In fact chemists have agreed to take 6×10^{23} as a standard number of particles—atoms, ions, molecules, electrons. The reason for the choice of this number is that one gramme of hydrogen, the lightest element, contains 6×10^{23} atoms. This number of particles is called one *mole*. Thus

6×10^{23} atoms of hydrogen is 1 *mole* of hydrogen atoms,
6×10^{23} molecules of water is 1 *mole* of water molecules,
6×10^{23} electrons is 1 *mole* of electrons,
and so on.

Part IV shows how, by the use of tables of *relative atomic mass* and a simple sum, the mass of one mole of any compound can be calculated. Having done this, measuring out one mole of a compound requires no more than a weighing machine.

(e) Moles and equations

The equation

$$SO_2(g) + 2H_2S(g) \longrightarrow 2H_2O(l) + 3S(s)$$

indicates that one molecule of SO_2 will react with 2 molecules of H_2S to form 2 molecules of water and 3 atoms of sulphur.

sulphur dioxide	+ hydrogen sulphide	→ water	+ sulphur
1 molecule	+ 2 molecules	→ 2 molecules	+ 3 atoms

Hence:

$$1 \times (6 \times 10^{23}) \text{ molecules} + 2 \times (6 \times 10^{23}) \text{ molecules} \rightarrow 2 \times (6 \times 10^{23}) \text{ molecules} + 3 \times (6 \times 10^{23}) \text{ molecules}$$

Since (6×10^{23}) particles is *1 mole*, we can write

1 mole of SO_2 molecules + 2 moles of H_2S molecules → 2 moles of H_2O molecules + 3 moles of S atoms

The number of particles represented in an equation can therefore tell us

(i) the number of atoms, molecules, ions or electrons involved,

or more usefully

(ii) the number of *moles* of atoms, molecules, ions or electrons involved.

For example

$$2H_2(g) + O_2(g) \longrightarrow 2H_2O(l)$$

means that 2 moles of hydrogen molecules react with 1 mole of oxygen molecules to form 2 moles of water molecules.

$$Al \longrightarrow Al^{3+} + 3e$$

means that 1 mole of aluminium atoms will ionise to form 1 mole of aluminium ions and 3 moles of electrons.

4.2 TECHNIQUES

Parts II and III of this book, dealing with chemical reactions and the behaviour of specific chemicals, are concerned with the experimentally observed facts of chemistry. This section considers the common apparatus and techniques behind these experimental facts.

(a) Safety
Chemical experiments should be undertaken only with the active supervision of a qualified chemist. For most experiments, and certainly all experiments involving acid, alkali or heat, safety spectacles should be worn. Skin contact with chemicals should be avoided, and for many reactions a fume cupboard is essential in order to avoid breathing in poisonous gases, fine sprays or powders.

(b) Common apparatus
Fig. 4.1 shows some standard laboratory equipment. For clarity, diagrams of apparatus in experimental reports are usually drawn as flat line drawings. This procedure is adopted in the remainder of this book.

(c) Purification techniques

(i) *Purity*

If a chemical is pure it may be one single element or one single chemical compound. A *mixture* contains at least two different pure chemicals. Mixtures can be *heterogeneous* or *homogeneous*.

The composition of a *heterogeneous* mixture is uneven. For example a fast-flowing river is a heterogeneous mixture of suspended sand and

fig 4.1 *standard laboratory equipment*

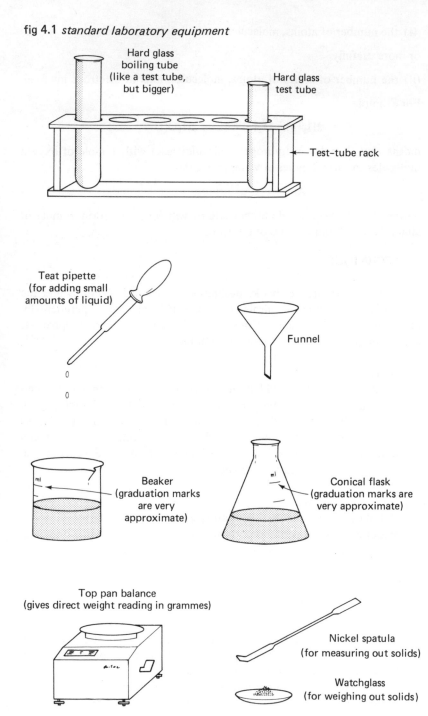

Hard glass
boiling tube
(like a test tube,
but bigger)

Hard glass
test tube

Test–tube rack

Teat pipette
(for adding small
amounts of liquid)

Funnel

Beaker
(graduation marks
are very
approximate)

Conical flask
(graduation marks are
very approximate)

Top pan balance
(gives direct weight reading in grammes)

Nickel spatula
(for measuring out solids)

Watchglass
(for weighing out solids)

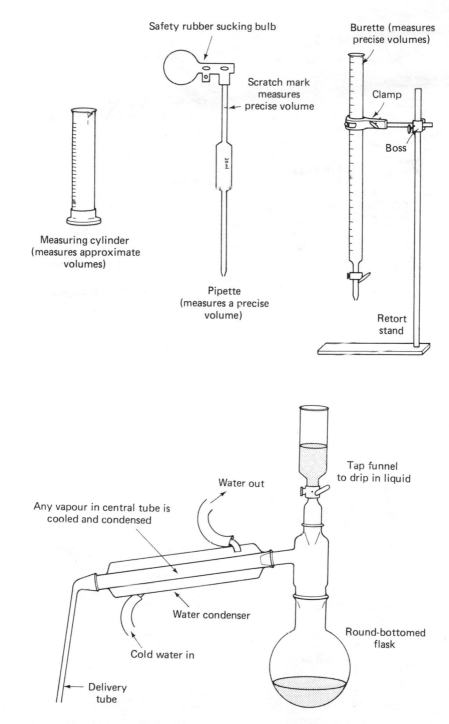

Safety rubber sucking bulb

Burette (measures precise volumes)

Scratch mark measures precise volume

Clamp

Boss

20 ml

Measuring cylinder (measures approximate volumes)

Pipette (measures a precise volume)

Retort stand

Water out

Tap funnel to drip in liquid

Any vapour in central tube is cooled and condensed

Water condenser

Round-bottomed flask

Cold water in

Delivery tube

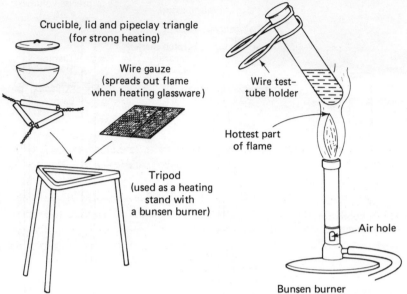

Crucible, lid and pipeclay triangle
(for strong heating)

Wire gauze
(spreads out flame
when heating glassware)

Wire test-
tube holder

Hottest part
of flame

Tripod
(used as a heating
stand with
a bunsen burner)

Air hole

Bunsen burner
(air hole open — hot roaring blue flame;
air hole closed — flickering yellow flame)

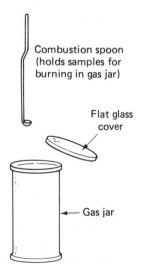

Combustion spoon
(holds samples for
burning in gas jar)

Flat glass
cover

Tongs (for heating
objects in a flame)

Gas jar

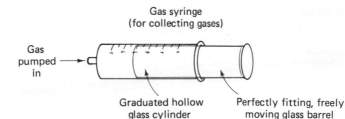

Gas syringe
(for collecting gases)

Gas
pumped
in

Graduated hollow
glass cylinder

Perfectly fitting, freely
moving glass barrel

mud, weed and water. Similarly, a fruit cake is heterogeneous, with a composition varying from raisin to cherry to cake depending upon which bit is being examined.

A *homogeneous* mixture has an even composition throughout. Air is a homogeneous mixture of nitrogen, oxygen, carbon dioxide and other gases. If sugar is stirred into warm water a homogeneous solution of sugar is obtained in which small samples, taken from any part of the mixture, will have the same sugar–water composition.

(ii) *Separating heterogeneous mixtures*

Often heterogeneous mixtures can be separated by physically pulling one of the chemicals away from the others. A mixture of sand and salt granules could be laboriously separated in this way with a fine pair of tweezers. A better way is suggested below.

A heterogeneous mixture of liquid and solid is conveniently separated by *filtration*. By stirring water with the sand–salt mixture above, solid sand may be separated by this technique from the resulting salt solution (Fig. 4.2). The separated solid is called the *residue*; the liquid dripping through is the *filtrate*.

(iii) *Separating homogeneous mixtures*

Homogeneous mixtures are commonly separated by *crystallisation, distillation* or *chromatography*.

(1) Crystallisation

This method can be used to separate pure salt from a solution of salt in water (see Fig. 4.3). The aqueous salt solution is heated in an evaporating dish in order to drive away some of the water as steam. In this way the concentration of the solution steadily increases.

The hot solution is periodically tested by allowing one drop to cool on the end of a glass rod. The formation of tiny crystals within the drop is an indication that crystals will also form if the main solution is allowed

fig 4.2 *filtration*

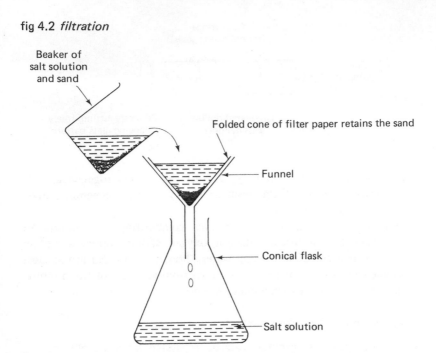

Beaker of
salt solution
and sand

Folded cone of filter paper retains the sand

Funnel

Conical flask

Salt solution

to cool. At this point heating is stopped and the dish set aside to cool slowly.

Slow cooling gives rise to large crystals, which can be filtered off as described above and dried by gently pressing between filter papers.

fig 4.3 *crystallisation*

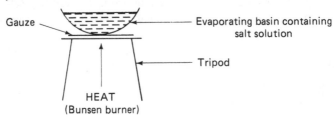

Gauze

Evaporating basin containing
salt solution

Tripod

HEAT
(Bunsen burner)

(2a) Distillation

Distillation separates a liquid from dissolved non-volatile solids. Pure water could be obtained from salt solution by this means (Fig. 4.4). The mixture is heated in the round-bottomed flask with a gentle bunsen flame, so that it boils steadily. Vapour forms on the *anti-bumping granules* (small pieces of stone). If these were left out the solution might *bump* (vaporise in sudden eruptions) with salt solution spluttering into the condenser.

53

fig 4.4 *distillation*

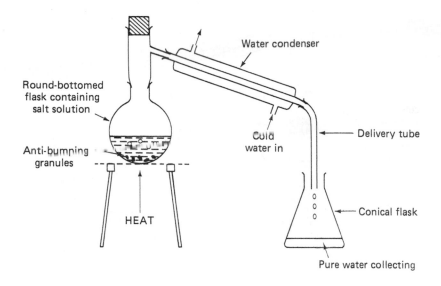

In the condenser, vapour is cooled and condenses back to liquid. This is the distillate, which is collected in the conical flask. The non-volatile salt, since it cannot vaporise, remains in the round-bottomed flask.

(2b) Fractional distillation

If a mixture of two liquids was heated in the distillation apparatus of Fig. 4.4, then both liquids would vaporise and distil into the collection flask together. Thus a mixture of water and alcohol could not be separated in this way. The distillate would contain a slightly increased concentration of whichever liquid has the lower boiling point (in this case the alcohol), but complete separation would not be achieved. In order to obtain pure alcohol, repeated collection and redistillation of the distillate would be necessary, with the concentration of the alcohol increasing slightly each time around. This process of continually condensing and revaporising occurs automatically on the inside of a *fractionating column*. This is no more than a long vertical tube packed with glass beads or small plates to provide a large surface upon which vapour can condense and revaporise. More-or-less pure alcohol could be separated from the mixture in a single distillation by employing a fractionating column (Fig. 4.5).

54

fiy **4.5** *fractional distillation*

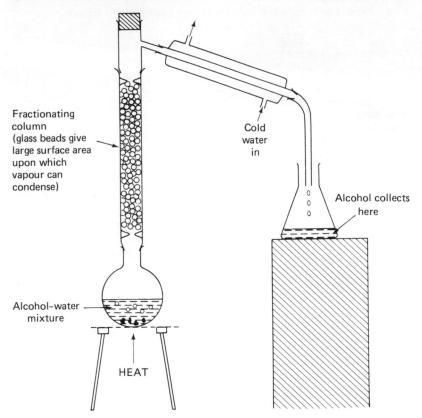

Fractionating column (glass beads give large surface area upon which vapour can condense)

Cold water in

Alcohol collects here

Alcohol-water mixture

HEAT

Fractional distillation is the technique used commercially to separate nitrogen and oxygen from liquid air, and to separate the many useful products from crude oil.

(3) Chromatography

Chromatography is used to separate complex chemicals such as proteins, dyes and vitamins. The use of chromatography to separate the complex mixture of dyes in ink illustrates the technique (Fig. 4.6).

The solvent steadily soaks into the paper, travelling up to the top. Some of the coloured dyes in the ink dot dissolve better than others as the solvent passes, and these get carried further up the paper. After the paper has dried, the different blobs of colour can be cut out with scissors—they have been separated from one another.

fig **4.6** *separating ink colours by chromatography: (a) arrangement of experiment, (b) a typical result after separation of colours*

Absorbent filter (or chromatography) paper

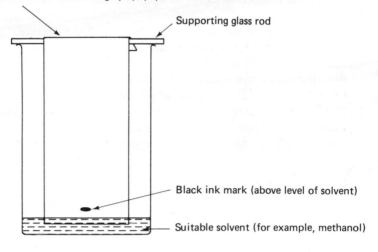

Supporting glass rod

Black ink mark (above level of solvent)

Suitable solvent (for example, methanol)

(a)

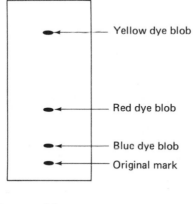

Yellow dye blob

Red dye blob

Blue dye blob

Original mark

(b)

SUMMARY OF CHAPTER 4

Formulae are determined experimentally.

Often formulae can be predicted from a knowledge of the charges on the ions concerned (ionic compounds) or the number of electrons commonly shared by the atoms concerned (covalent compounds).

The chemicals involved in a reaction may be represented by symbols in a chemical equation. Numbers are placed in front of the symbols such that the number of atoms of each element is the same before and after reaction.

One mole of substance is 6×10^{23} particles of that substance.

Mixtures of even composition are homogeneous, those of uneven composition are heterogeneous.

Filtration separates a solid from a liquid.

Crystallisation separates a dissolved solid from solution.

Distillation separates a liquid from dissolved non-volatile substance.

Fractional distillation separates a homogeneous mixture of liquids.

Chromatography separates complex molecules.

PART II
CHEMICAL REACTIONS

Electrons bond atoms together. In metallic bonds these electrons are delocalised, in covalent bonding they are shared, and in ionic bonding they have jumped from one atom to another; in each case it is the electrons which are responsible for the bond. When chemicals react, old bonds break and new bonds form, and therefore chemical reactions are no more than movements of electrons. This part looks at the ways in which electrons move between atoms, and what happens when they do.

CHAPTER 5

ENERGY CHANGES IN CHEMISTRY

A small lump of potassium, when dropped into water, rushes about on the surface as a molten ball of metal, sparking, bursting into flame, hissing and heating up the water as it goes.

Magnesium burns in air to give out a brilliant light and a large quantity of heat.

When a mixture of hydrogen and oxygen gases are sparked, there is a deafening bang.

Although most chemical reactions are not quite so spectacular, any reaction is accompanied by *some* energy change. The above examples involve the production of heat, light and sound energy.

In contrast some reactions *take in* energy as they occur. For example, carbon disulphide, a foul-smelling poisonous liquid used in the manufacture of synthetic fibres, is obtained industrially by forcing carbon and sulphur to combine together. However, this reaction only occurs if energy is pushed into the two reacting substances, most conveniently by putting in heat. The more trivial 'reaction' in which liquid held on the hand feels cold as it evaporates results from the liquid *taking* heat energy from the hand to allow it to vaporise.

5.1 AGREED STANDARDS

If the amount of energy given out or taken in by various chemical reactions is to be compared, then a standard way of measuring this energy must be agreed.

(i) *Heat* should be the *only* form of energy involved in the reaction. Thus magnesium would be burned in oxygen in a completely sealed container in order that the brilliant light could not escape through the walls except as heat. Similarly hydrogen and oxygen would be exploded in a strong steel box immersed in water. The shock waves of the explosion then become absorbed by the surrounding water as heat rather than escape as sound.

(ii) The *mole* (see Section 4.1(d)) has been chosen as the standard amount of chemical involved in the reaction.

Heat changes of molar quantities are given the internationally agreed symbol ΔH, and this heat energy is measured in joules (abbreviated to J) or more conveniently in thousands of joules, kilojoules (kJ). To give some idea of this energy unit, a match, when struck, gives out about 10 kJ of energy.

A reaction *giving out heat* is described as an *exothermic* reaction. Since heat is lost by exothermic reactions, their heat change, or ΔH value, is given a *negative* sign. For example, if 1 mole (24 g) of magnesium metal is burned in air, an exothermic reaction occurs in which 600 kJ of heat are given out (equivalent to burning sixty matches simultaneously). For this reaction $\Delta H = -600$ kJ per mole of magnesium, or -600 kJ mol^{-1}.*

A reaction *taking in heat* is described as an *endothermic* reaction. Here the chemicals are gaining heat energy and ΔH takes a *positive* sign. For example, when 1 mole (12 g) of carbon reacts with sulphur, a net 88 kJ of heat have to be supplied. For this reaction $\Delta H = +88$ kJ per mole of carbon, or $+88$ kJ mol^{-1}.*

5.2 DIFFERENT ΔH VALUES

The following ΔH values are common:

(i) the *heat of combustion*, ΔH_c, is the heat change when 1 mole of substance burns completely in oxygen;

(ii) the *heat of formation*, ΔH_f, is the heat change when 1 mole of substance is formed from its constituent elements in their normal state;

(iii) the *heat of solution*, ΔH_s, is the heat change when 1 mole of substance dissolves completely in water;

(iv) the *heat of reaction*, ΔH_r, is the heat change when the number of moles appearing in the equation do react;

(v) the *heat of melting* (or heat of fusion), ΔH_m, is the heat change when 1 mole of substance is melted.

Examples of various ΔH values are given below.

(a) Heat of combustion, ΔH_c

A bunsen burner mixes methane, the major constituent of our domestic gas supply, with air, and then burns this mixture. In books of chemical data, ΔH_c for methane is given as -890 kJ mol^{-1}. This means that when 1 mole of methane burns completely (the average bunsen burner would

---

*mol^{-1} simply means 'per mole'. Thus kJ mol^{-1} means 'kilojoules per mole'.

burn this amount in about thirty minutes), 890 kJ of heat energy are *given out*. Methane burns completely to form carbon dioxide and water, as shown in the following equation

$$CH_4(g) + 2O_2(g) \longrightarrow CO_2(g) + 2H_2O(l) \quad \Delta H_c = -890 \text{ kJ mol}^{-1}$$
methane oxygen   carbon   water
                      dioxide

All chemicals containing carbon and hydrogen form these same two products on complete combustion in oxygen or air. For example, the reaction for propane gas, C_3H_8, is

$$C_3H_8(g) + 5O_2(g) \longrightarrow 3CO_2(g) + 4H_2O(l) \quad \Delta H_c = -2200 \text{ kJ mol}^{-1}$$

and that for ethanol (alcohol), C_2H_6O, is

$$C_2H_6O(l) + 3O_2(g) \longrightarrow 2CO_2(g) + 3H_2O(l) \quad \Delta H_c = -1400 \text{ kJ mol}^{-1}$$

(b) Heat of formation, ΔH_f

A chemical compound can be made in many different ways. For example carbon dioxide is obtained by
(i) heating calcium carbonate

$$CaCO_3(s) \longrightarrow CaO(s) + CO_2(g)$$

(ii) adding hydrochloric acid to sodium carbonate solution

$$2HCl(aq) + Na_2CO_3(aq) \longrightarrow 2NaCl(aq) + H_2O(l) + CO_2(g)$$

(iii) burning carbon in oxygen gas

$$C(s) + O_2(g) \longrightarrow CO_2(g)$$

The heat involved in each reaction will be different, yet carbon dioxide is formed in each case. It has been agreed that heat of formation, ΔH_f, should refer only to the formation of one mole of compound from its elements as they normally exist. Thus only the energy change in reaction (iii) gives ΔH_f, because here carbon dioxide is formed from its constituent elements in normal state, solid carbon and oxygen gas.
    Further examples of ΔH_f are given below.

$$S(s) + O_2(g) \longrightarrow SO_2(g) \quad \Delta H_f = -297 \text{ kJ mol}^{-1}$$

$$Mg(s) + Cl_2(g) \longrightarrow MgCl_2(s) \quad \Delta H_f = -642 \text{ kJ mol}^{-1}$$

$$H_2(g) + \tfrac{1}{2}O_2(g) \longrightarrow H_2O(l) \quad \Delta H_f = -286 \text{ kJ mol}^{-1}$$

(c) Heat of solution, ΔH_s

If common salt, sodium chloride, is shaken in a test tube of water, the tube begins to get noticeably colder as the salt dissolves; heat energy is being taken from the water and tube. This means that energy is *required* for salt to dissolve in water. Thus ΔH_s for sodium chloride will be positive (heat taken in). In fact, data books give the value $\Delta H_s = +9$ kJ mol^{-1} for sodium chloride. This means that if 1 mole of sodium chloride (58.5 g) is shaken with sufficient water, 9 kJ of energy are taken from the water.

In the simple process of dissolving, a substance remains chemically unchanged. Salt remains as salt and sugar remains as sugar even once they have dissolved. Equations showing a substance dissolving in water at first glance look rather trivial, simply showing the change of situation of the chemical. In such equations a large quantity of water is represented by the symbol 'Aq':

$$NaCl(s) + Aq \longrightarrow NaCl(aq) \quad \Delta H_s = +9 \text{ kJ mol}^{-1}$$

One mole of potassium iodide, KI, is found to require even more energy to dissolve

$$KI(s) + Aq \longrightarrow KI(aq) \quad \Delta H_s = +31 \text{ kJ mol}^{-1}$$

and when sodium thiosulphate (photographic hypo), $Na_2S_2O_3$, is shaken in water, the mixture becomes exceptionally cold because still more energy is required

$$Na_2S_2O_3(s) + Aq \longrightarrow Na_2S_2O_3(aq) \quad \Delta H_s = +48 \text{ kJ mol}^{-1}$$

(d) Heat of reaction, ΔH_r
Data books tell us that for the industrial reaction

$$N_2(g) \quad + \quad 3H_2(g) \longrightarrow 2NH_3(g)$$
$$\text{nitrogen} \quad \text{hydrogen} \quad \text{ammonia}$$

$\Delta H_r = -92$ kJ mol^{-1}. This means that when 1 mole of nitrogen gas reacts with 3 moles of hydrogen gas (these are the quantities shown in the equation), 92 kJ of heat will be given out.

Similarly

$$2Ag(s) + Cl_2(g) \longrightarrow 2AgCl(s) \quad \Delta H_r = -254 \text{ kJ mol}^{-1}$$

This means that when 2 moles of silver react with 1 mole of chlorine (the quantities stated in the equation), 254 kJ of heat are given out.

Experiments and the calculations which determine these various ΔH values will be met in Part IV of this book on Chemical Calculations (see Section 16.6).

5.3 REASONS FOR ENERGY CHANGES

To *break* a bond, *energy will have to be put in* to pull the atoms apart. To *form* a bond, *energy will be given out.*

All reactions involve either the breaking of bonds, or the making of new bonds, or both. The energy changes involved can be represented in energy diagrams in which energy provided (breaking bonds) is shown by a move up the page (Fig. 5.1) and energy given out (forming bonds) by a move down the page (Fig. 5.2).

fig 5.1 *breaking a bond*

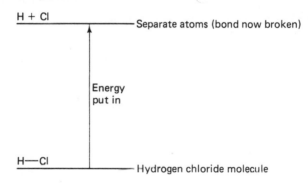

fig 5.2 *forming a bond*

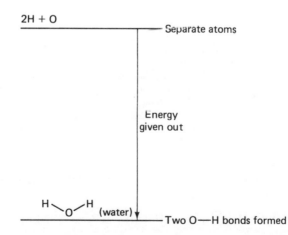

These diagrams are useful in showing the energy changes of a complete reaction. For example the reaction

64

$$H_2(g) \ + \ Cl_2(g) \ \longrightarrow \ 2HCl(g)$$

or

$$H-H \ + \ Cl-Cl \ \longrightarrow \ \begin{array}{c} H-Cl \\ H-Cl \end{array}$$

Here the H–H and Cl–Cl bonds have been broken (energy required) and two H–Cl bonds have been formed (energy given out). Furthermore, experiment shows that the net result is an exothermic reaction (energy is given out, ΔH_r is negative) indicating that the energy gained when H–Cl bonds are formed must be greater than that needed when the H–H and Cl–Cl bonds are broken (see Fig. 5.3).

fig 5.3 *the reaction of hydrogen with chlorine*

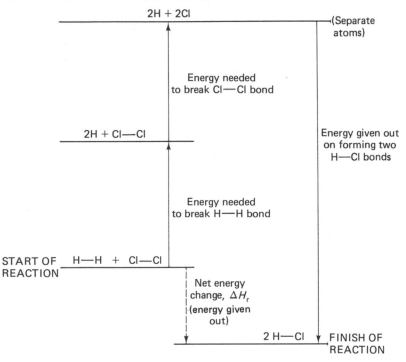

Energy changes involved in the formation of water from hydrogen and oxygen gases can be represented similarly:

$$2H_2(g) + O_2(g) \ \longrightarrow \ 2H_2O(l) \quad \Delta H_r = -572 \ kJ \ mol^{-1}$$

Experiment shows the net energy change, ΔH_r, to be negative, suggesting that more energy is given out when the O–H bonds are formed than is required to break H–H and O=O bonds (see Fig. 5.4).

fig 5.4 *the reaction of hydrogen with oxygen*

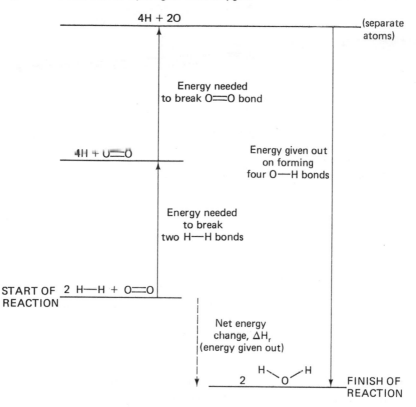

5.4 BOND STRENGTHS

Strong bonds will need more energy to break them up than weak bonds, and will give out more energy when they are formed. For example nitrogen molecules contain a very strong bond, whereas the bond in fluorine molecules is exceptionally weak (one reason why fluorine is so reactive and dangerous, see Section 12.1) (see Fig. 5.5).

5.5 IONIC BONDS

The examples given so far concern the breaking and formation of covalent bonds. In exactly the same manner the breaking of an ionic bond requires energy whilst the formation of the bond gives out energy. It is for this reason that sodium chloride takes in energy as it dissolves, making the water colder (see Section 5.2(c)). The ionic bonds between an Na^+ and a Cl^- ion have to be broken (i.e. the ions have to be pulled apart) before

fig **5.5** *comparison of bond strengths of N$_2$ and F$_2$*

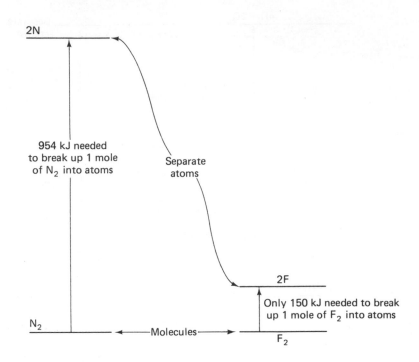

they can freely move about in solution as dissolved ions.

In contrast, when barium chloride solution is mixed with sodium sulphate solution, solid lumps of barium sulphate are produced instantaneously in a strongly exothermic reaction. The release of energy is explained by the *formation* of ionic bonds between the barium and the sulphate ions:

$$Ba^{2+}(aq) + SO_4^{2-}(aq) \longrightarrow Ba^{2+}SO_4^{2-}(s)$$

in barium chloride solution   in sodium sulphate solution   the ions are now ionically bonded in solid barium sulphate

5.6 WHY REACTIONS HAPPEN

Many everyday chemical reactions are 'one way'. For example a match is struck, heat and light are given out and the match burns (Fig. 5.6). Giving back the energy by heating up a dead match and shining a light on it will not reverse the process. Likewise, running a car backwards down a hill

fig **5.6** *energy change when a match is struck*

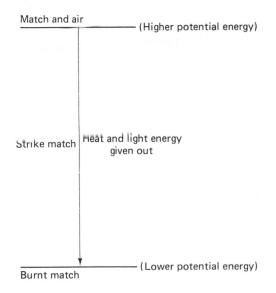

does not make it suck in fumes through the exhaust pipe, manufacturing petrol as it goes.

Are there any rules to determine whether a particular reaction should or should not occur?

In the majority of reactions, stored (potential) energy is given out. The chemicals give up potential energy to become more stable (Fig. 5.6).

Perhaps, then, only those reactions which give out energy can occur? This is certainly believed to be an important factor, but not the only factor, because there are a few spontaneous reactions which actually *take in* energy. For example, a few drops of the liquid ether on the hand evaporate very quickly leaving the hand feeling cold. The ether has *taken in* heat energy from the hand during evaporation (Fig. 5.7). In addition to now holding more energy, the molecules of ether are, after evaporation, more randomly spread out. They have a greater degree of disorder (Fig. 5.8).

It has been suggested that, apart from energy considerations, a reaction is favoured if the constituents can become more disordered. The degree of disorder of particles is called the *entropy*. The more disordered and spread out the particles of a reaction become, the greater the entropy and the more likely that the reaction should occur.

Whether a reaction could happen or not therefore depends upon two factors, as shown in Table 5.1.

fig 5.7 *energy change when ether evaporates*

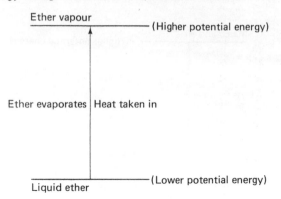

fig 5.8 *the increase in randomness when ether evaporates*

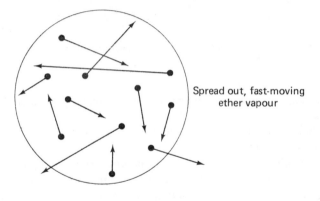

Spread out, fast-moving ether vapour

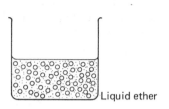

Liquid ether

SUMMARY OF CHAPTER 5

Chemical reactions always give out or take in energy.

Energy changes are most conveniently measured by making sure that they occur only as heat changes.

Table 5.1 *factors which determine whether a reaction happens*

(1) *The energy factor*

Is energy given out?	Yes	Yes	No	No

(2) *The entropy factor*

Do the particles become more random?	Yes	No	Yes	No

Here reaction can happen. For example, when a candle burns, heat energy is given out and the particles become more random as solid candle wax becomes carbon dioxide gas and water vapour.

Reaction can happen only if so much energy is given out that the entropy factor is outweighed. For example, the exothermic reaction

$$2H_2S(g) + SO_2(g) \longrightarrow 2H_2O(l) + 3S(s)$$

Here the particles become less random (from gases to a solid and a liquid), but much energy is given out.

Reaction can happen if the increase in disorder more than outweighs the energy taken in. For example, when sugar crystals dissolve in water the mixture becomes colder as the sugar molecules take in energy to break apart from one another. But in so doing, their disorder increases considerably (from solid crystal to dissolved molecules).

In this case, reaction cannot happen.

Reactions giving out heat are *exothermic*, those taking in heat are *endothermic*.

Heat changes of reactions involving molar quantities are given the symbol ΔH.

For exothermic reactions ΔH is negative, for endothermic reactions it is positive.

Energy is required (taken in) to break bonds; energy is therefore given out when bonds are formed.

A reaction is favoured if (1) energy is given out and/or if (2) entropy (randomness) increases.

CHAPTER 6

THE SPEED AND DIRECTION OF CHEMICAL REACTIONS

6.1 RATE OF REACTION

Chemical reactions can be exceptionally fast. Lighting gun powder, sparking a petrol-air mixture or striking a match provide examples of almost instantaneous reactions. Others can be quite slow. The chemical reaction that converts sweet grape juice into alcohol takes several weeks, the souring of milk one or two days. Epoxy resin adhesives such as 'Araldite' take several hours to harden. Reactions of substances called 'linoleic esters' take many years to bring about the hardening of oils in an oil painting, and although theoretically diamonds are steadily changing to graphite, the infinitely long reaction time for this conversion allows the owners of these gems to sleep soundly at night.

(a) Reaction rate in society
Many chemical reactions that are potentially useful to mankind were initially too slow. Those producing polythene, Teflon, synthetic rubber, fertilisers, flavourings and industrial solvents were all too slow to be of any use when first discovered. By understanding the factors affecting reaction speed, Man has been able to choose the conditions under which each of these industrial reactions is viable.

In contrast, particularly in the food industry, an understanding of reaction rate has enabled harmful processes such as food decay to be slowed down appreciably.

(b) Measuring rates of reaction
Experiments measuring the rate of reaction under different conditions show us what factors do affect rate. In order to see how the rate of a reaction alters, it is necessary to choose any change which occurs in that reaction as it progresses. By following this change, we can thus follow the reaction rate. This can be best understood by looking at a number of examples.

(i) *The reaction between calcium carbonate and hydrochloric acid*

Chalk, limestone and marble are all forms of calcium carbonate which react with hydrochloric acid according to

$$CaCO_3(s) + 2HCl(aq) \longrightarrow CaCl_2(aq) + H_2O(l) + CO_2(g)$$

| solid chalk | hydrochloric acid | calcium chloride solution | water | carbon dioxide |

As this reaction proceeds, an increasing volume of carbon dioxide gas is obtained. This can be collected in a gas syringe and the volume of gas collected measured every half minute. A fast increase in the volume of gas collected would indicate a fast reaction, and vice versa.

The apparatus for doing this is shown in Fig. 6.1. The test tube is carefully balanced inside the flask before the reaction is started in order to keep chalk and acid apart. To start the reaction, the flask is shaken, and then the stop clock is started.

fig 6.1 *measuring reaction rate of hydrochloric acid with calcium carbonate*

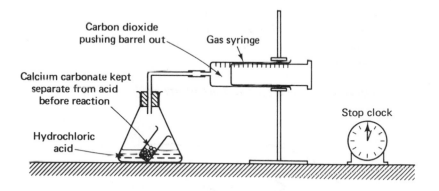

From this experiment, a graph of time from the start against volume of gas collected has the shape shown in Fig. 6.2. *The steeper the line* the more gas must be collecting in a given time and therefore *the faster the reaction*. Thus the reaction is fast at the start (steep slope at O), slowing down as the reaction progresses (steepness falls off). Indeed it would be expected that the reaction should slow down as the chemicals gradually become used up.

fig 6.2 *volume of carbon dioxide collected in reaction of hydrochloric acid with chalk*

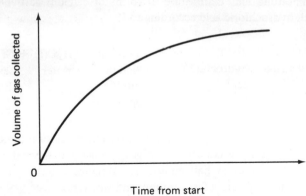

Graphs obtained under different conditions from those used in this experiment are now drawn in the following three cases.

(1) Varying the particle size of the chalk

In these experiments, temperature, mass of chalk and concentration and volume of acid were all kept the same. Each of the three curves shown on Fig. 6.3 corresponds to an experiment using different sizes of chalk particles, ranging from large lumps to powder.

Conclusion from Fig. 6.3: reaction rate is increased by reducing particle size (the smallest particles—powder—show the steepest slope and thus the fastest reaction).

fig 6.3 *effect of varying the particle size of the chalk*

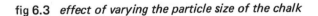

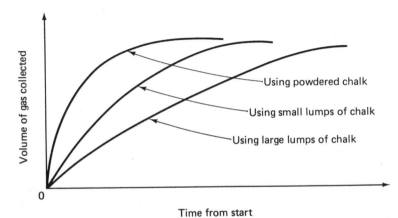

(2) Varying the concentration of the hydrochloric acid

In these experiments, temperature, mass and particle size of chalk and volume of acid were all kept the same. The results are shown as three curves in Fig. 6.4.

Conclusion from Fig. 6.4: reaction rate is increased by increasing the concentration (the greatest concentration of acid gave the steepest slope).

fig 6.4 *effect of varying the concentration of the acid*

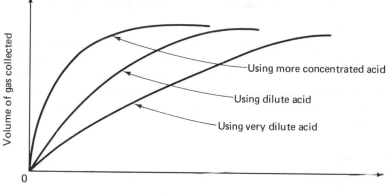

(3) Varying the temperature

The apparatus of the previous experiments would be used, but with the flask submerged in a large tank of water maintained at various fixed temperatures. In these experiments, the mass and particle size of the chalk

fig 6.5 *effect of varying the temperature at which the reaction takes place*

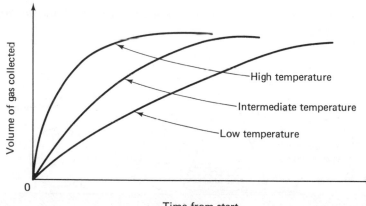

and the concentration and volume of acid were all kept the same. Fig. 6.5 shows the results.

Conclusion from Fig. 6.5: reaction rate is increased by increasing the temperature.

(ii) *The reaction between magnesium and hydrochloric acid*

The reaction can be written as

$$Mg(s) + 2HCl(aq) \longrightarrow MgCl_2(aq) + H_2(g)$$

magnesium   hydrochloric   magnesium   hydrogen
           acid   chloride

Since a gas is again given off, this reaction could be followed using the same apparatus as for reaction (i). In this case, it would be the volume of hydrogen gas that would be measured each half minute, but the general form of the graphs and the conclusions drawn from them would be exactly as described in (i).

(iii) *The reaction between sodium thiosulphate and hydrochloric acid*

$$Na_2S_2O_3(aq) + 2HCl(aq) \longrightarrow 2NaCl(aq) + H_2O(l) + SO_2(g) + S(s)$$

sodium   hydrochloric   sodium   water   sulphur   solid
thiosulphate   acid   chloride   dioxide   sulphur

In this reaction, the fine suspension of tiny solid sulphur particles, $S(s)$, increases as the reaction proceeds, giving a more and more cloudy liquid. So, for this reaction, the faster the cloudiness appears the greater the rate of reaction.

A means of following reaction rate by making use of this cloudiness is illustrated in Fig. 6.6. The reaction is started by pouring a measured quantity of sodium thiosulphate solution into hydrochloric acid in the

fig 6.6 *measuring reaction rate of hydrochloric acid with sodium thiosulphate*

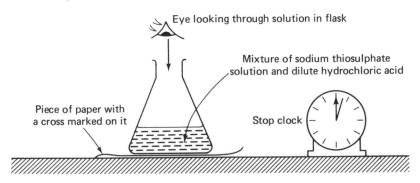

flask, simultaneously starting a stop clock. Measuring the time taken for the cloudiness to obscure the cross under various different conditions enables the rates of reaction for these conditions to be compared. In order to make a fair comparison when altering acid or thiosulphate concentrations, all other factors (concentrations of other reagent, temperature, total volume of liquid in the flask) would have to be kept the same.

Table 6.1 summarises the results and conclusions obtained from such experiments.

Table 6.1 *results of rate experiments in the reaction of sodium thio-sulphate with an acid*

Experiments	Result	Conclusion
Using different concentrations of sodium thiosulphate solution	cross obscured faster with high concentrations	rate increases with increased concentration
Using different concentrations of hydrochloric acid	cross obscured faster with high concentrations	rate increases with increased concentration
Carrying out the same experiment at different temperatures	cross obscured faster at higher temperatures	rate increases with increased temperature

(c) Catalysts

Some reactions speed up dramatically in the presence of one specific chemical, even though the chemical itself is not used up in the reaction.

A chemical that alters the rate of a reaction without itself being permanently consumed is called a *catalyst*.

For example, a flask of hydrogen peroxide solution, $H_2O_2(aq)$ (used in some bleaches and mouthwashes), very slowly breaks down into water and oxygen gas when left at room temperature:

$$2H_2O_2(aq) \longrightarrow 2H_2O(l) + O_2(g)$$

If a few grains of black copper(II) oxide are added, oxygen gas can be seen to bubble out in a steady stream. Or if the same amount of black grains of manganese(IV) oxide are added, clouds of oxygen gas pour from the solution, frequently causing it to 'boil over'. The masses of these oxides remain unaltered at the end of the reaction, showing that copper(II) oxide and manganese(IV) oxide both *catalyse* this reaction: they are both *catalysts*.

More examples of catalysts are now given.

(i) *The reaction of nitrogen with hydrogen*

$$N_2(g) + 3H_2(g) \longrightarrow 2NH_3(g)$$
nitrogen   hydrogen   ammonia

This reaction is the basis of fertiliser manufacture. Economic quantities of ammonia can only be obtained if the reaction is carried out in the presence of *fine iron filings*, which catalyse the process.

(ii) *The reaction of sulphur dioxide with oxygen*

$$2SO_2(g) + O_2(g) \longrightarrow 2SO_3(g)$$
sulphur   oxygen   sulphur
dioxide   trioxide

This reaction provides the first stage in the industrial production of concentrated sulphuric acid, which itself is used in making paints, dyes, plastics, steel and synthetic fibres. Reaction rate is exceptionally low, making the process uneconomic unless a catalyst of vanadium(V) oxide, V_2O_5, is added.

(iii) *The reaction of ammonia with oxygen*

$$4NH_3(g) + 5O_2(g) \longrightarrow 6H_2O(l) + 4NO(g)$$
ammonia   oxygen   water   nitrogen
monoxide

This reaction rate is undetectable in the absence of a catalyst. However, in the presence of platinum metal, it is explosive (see Fig. 6.7).

fig 6.7 *catalysing the reaction of oxygen and ammonia*

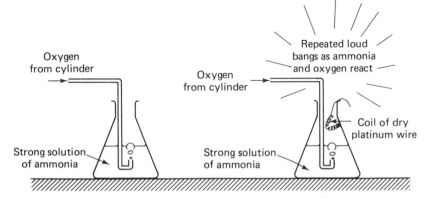

Oxygen from cylinder

Oxygen from cylinder

Repeated loud bangs as ammonia and oxygen react

Coil of dry platinum wire

Strong solution of ammonia

Strong solution of ammonia

(d) The factors affecting reaction rate
For a reaction to occur, individual particles of the chemicals concerned

must *collide*. Furthermore, if collision is to lead to a reaction, they must collide with a certain minimum amount of energy, called *activation energy*. With this in mind, explanations can be found for the influences upon rate that have just been observed.

(i) *Rate increases with concentration*

A more concentrated solution contains more particles in a given volume. Thus more collisions are likely between particles, giving rise to an increase in reaction rate.

(ii) *Rate increases with temperature*

As a liquid or a gas is heated, its particles move faster (they have more kinetic energy). This greater movement means (1) that there are more collisions at higher temperature, and therefore there is a faster rate of reaction, and (2) that at higher temperatures there are more particles moving fast and thus colliding hard enough to react.

(iii) *Rate increases with decrease in particle size*

The more that a lump of solid is split up, the more surface becomes exposed and the *greater the surface area* (Fig. 6.8). With a greater surface area, there is more likelihood that particles of other chemicals will collide with the solid, giving rise to a greater reaction rate.

fig 6.8 *breaking up a solid increases the surface area*

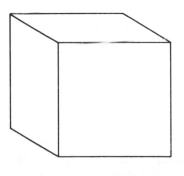 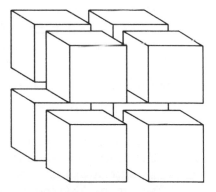

Lump of solid

Same lump of solid split up
(much greater surface exposed)

(iv) *Rate increases in the presence of a catalyst*

Many catalysts form a temporary bond with one of the chemicals in the reaction, and in so doing they weaken other bonds within that chemical. Thus bonds which need to be broken in the reaction can now be broken more readily, and the reaction can proceed more swiftly.

Other catalysts are temporarily used up, reacting to form a different compound. This new compound readily breaks up to form the required product whilst at the same time regenerating the catalyst. In this way the reaction is 'helped along an easier path' by the catalyst.

6.2 REVERSIBLE REACTIONS

Some reactions can be made to go in either direction simply by changing the conditions. A block of ice from the refrigerator takes in heat and slowly turns into water:

$$\text{Heat} + H_2O(s) \longrightarrow H_2O(l)$$
$$\phantom{\text{Heat} + }\text{ice} \text{water}$$

If replaced in the refrigerator, heat is given up by the water as it turns back into ice:

$$\text{Heat} + H_2O(s) \longleftarrow H_2O(l)$$

This is a *reversible reaction*: it can go in either direction. Reversible reactions are written using the symbol $\rightleftharpoons$ in order to show that the reaction can go either way:

$$\text{Heat} + H_2O(s) \rightleftharpoons H_2O(l)$$

Common salt and water give a salt solution when heated. On cooling, salt crystals reappear:

$$\text{Heat} + NaCl(s) \rightleftharpoons NaCl(aq)$$
$$\phantom{\text{Heat} + }\text{solid} \text{salt}$$
$$\phantom{\text{Heat} + }\text{salt} \text{solution}$$

Heating water produces steam. Cooling steam condenses back the water:

$$\text{Heat} + H_2O(l) \rightleftharpoons H_2O(g)$$
$$\phantom{\text{Heat} + }\text{water} \text{steam}$$

Certain chemical reactions too can be reversed. Crystals of copper(II) sulphate are blue. They have the formula $CuSO_4.5H_2O$, and contain water molecules attracted around the copper and sulphate ions. The crystals are therefore said to be *hydrated*. If these hydrated copper(II) sulphate crystals are heated in a test tube they turn white, forming *anhydrous* copper(II) sulphate as water is driven from them. If water is then added back to this white solid, blue hydrated crystals are reformed and the tube becomes very hot. The reactions can be written as

$$\text{Heat} + CuSO_4.5H_2O(s) \rightleftharpoons CuSO_4(s) + 5H_2O(l)$$

6.3 EQUILIBRIUM

Ethanol is the alcohol of drink. Ethanoic acid is vinegar. If mixed they react slowly together, forming ethylethanoate and water. Ethylethanoate is used commercially in many glues and varnishes. The reaction is

$$\text{ethanol} + \text{ethanoic acid} \longrightarrow \text{ethylethanoate} + \text{water}$$

As ethanol and ethanoic acid react away, so the concentration of both chemicals decreases and the rate of their reaction slows down. But this is not the end of the reaction, because the ethylethanoate and water that have just been formed slowly react to give ethanol and ethanoic acid again:

$$\text{ethanol} + \text{ethanoic acid} \longleftarrow \text{ethylethanoate} + \text{water}$$

As the concentrations of ethylethanoate and water build up, so the rate of this back reaction increases.

Starting with a mixture of ethanol and ethanoic acid:

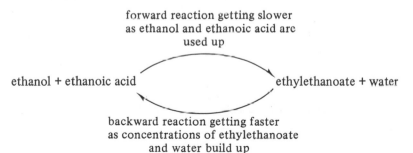

forward reaction getting slower
as ethanol and ethanoic acid are
used up

ethanol + ethanoic acid   ethylethanoate + water

backward reaction getting faster
as concentrations of ethylethanoate
and water build up

Thus the forward reaction slows down and the backward reaction speeds up, until eventually the rates of both reactions are equal. The reversible reaction is then in *equilibrium*. At equilibrium, each chemical is being produced by one reaction as fast as it is being used up by the reverse reaction. At equilibrium, the concentration of each chemical will therefore *remain constant*.

This is not to say that at equilibrium the two reactions stop. Reaction continues in both directions, but at the same rate. Ethanol, for example, will be reacting away as fast as it is being produced. This continuance of reaction at equilibrium is usually stressed by describing it as 'moving equilibrium' or *dynamic equilibrium*. Any reversible reaction, if left long enough, will reach a state of dynamic equilibrium. The above example requires several months at normal temperatures before constant equilibrium concentrations are reached. Of course, equilibrium would be reached sooner if both forward and backward reactions were speeded up by, for example, heating or adding a catalyst (see Section 6.1).

6.4 Le CHATELIER'S PRINCIPLE

Consider the same reaction as in the previous section:

$$\text{ethanol} + \text{ethanoic acid} \rightleftharpoons \text{ethylethanoate} + \text{water}$$

Mix in flask

1 mole of ethanol
+
1 mole of ethanoic acid

Three months later (chemicals now in dynamic equilibrium)

$\frac{1}{3}$ mole of ethanol
+
$\frac{1}{3}$ mole of ethanoic acid
+
$\frac{2}{3}$ mole of ethylethanoate
+
$\frac{2}{3}$ mole of water

In this particular equilibrium reaction, the formation of ethylethanoate and water seems to be favoured. Nevertheless one-third of the ethanol and ethanoic acid remains. Our picture of equilibrium can be used to find a means of improving the yield of ethylethanoate.

More ethylethanoate could be obtained by stopping the backward reaction which uses it up again:

$$\text{ethanol} + \text{ethanoic acid} \xrightleftharpoons{x} \text{ethylethanoate} + \text{water}$$

back
reaction
stopped

This could be achieved by adding concentrated sulphuric acid to the mixture. This acid forms very strong bonds with water; it is a strong dehydrating agent. By taking away the water, the ethylethanoate has nothing with which to react and the backward reaction is prevented. The forward reaction continues until virtually all the ethanol and ethanoic acid have reacted away.

In 1885 H. L. Le Chatelier suggested a useful rule for predicting such effects on chemicals in equilibrium. Le Chatelier's principle simply states that 'whatever is done to an equilibrium mixture, it will tend to do the opposite'. Thus for our reaction

$$\text{ethanol} + \text{ethanoic acid} \rightleftharpoons \text{ethylethanoate} + \text{water}$$

if water is removed from this equilibrium, the mixture will tend to do the opposite, and produce more water by moving more in this direction →.
Pessimists claim Le Chatelier's principle as a rule of life. Its application to industrial reactions is illustrated in the first two of the following examples.

(a) The industrial production of ammonia

A major function of fertilisers is to put nitrogen back into the soil. The Haber process uses the reversible reaction

$$N_2(g) + 3H_2(g) \rightleftharpoons 2NH_3(g)$$

to convert nitrogen of the air into ammonia, the vital ingredient of fertilisers. The gases in this process react in the presence of an iron catalyst. The yield of ammonia from the mixture is very low since the ammonia formed readily splits into nitrogen and hydrogen again by the reverse reaction. By applying Le Chatelier's principle to determine optimum temperature and pressure, the yield can be improved considerably.

(i) Temperature

The reaction producing ammonia is exothermic:

$$N_2(g) + 3H_2(g) \rightleftharpoons 2NH_3(g) + Heat$$

According to Le Chatelier's principle, removing heat would encourage the equilibrium mixture to do the opposite by producing heat. Thus the forward reaction, producing ammonia and heat would be favoured at low temperature. The lower the temperature the more ammonia we would expect to find in the equilibrium mixture. Of course, too low a temperature would mean that we may have to wait twenty years to reach equilibrium, so a compromise temperature of 500°C (quite cool by industrial standards) is used in practice. So we have

Low temperature = high yield of NH₃ eventually, but reaction too slow

High temperature = low yield of NH₃ obtained quickly

(ii) Pressure

The equation for the reaction shows that (1 + 3) = 4 moles of nitrogen and hydrogen react to produce only 2 moles of ammonia. The number of moles of a gas is proportional to the volume it occupies (see Section 15.5). Therefore 4 volumes of nitrogen and hydrogen would be reduced to just 2 volumes of ammonia. If pressure were increased the equilibrium mixture would move to do the opposite, relieving pressure by reducing its own volume. This volume reduction is achieved by nitrogen and hydrogen reacting to produce ammonia, so increasing the yield of ammonia. Increasing the pressure also has the advantage of increasing the concentration of the gases and thus increasing reaction rate. So now we have

High pressure = high yield of NH_3 *obtained quickly*

(b) The industrial production of sulphur trioxide
The Contact process manufactures sulphur trioxide from sulphur dioxide and oxygen in the air as the first stage in the large-scale production of sulphuric acid. The reaction is exothermic:

$$2SO_2(g) + O_2(g) \rightleftharpoons 2SO_3(g) + Heat$$

(i) *Temperature*

Le Chatelier's principle predicts that removing heat would persuade the reaction to do the opposite and produce more heat, and with it more sulphur trioxide. Again, however, too low a temperature means too slow a reaction. So we have

Low temperature = high yield of SO_3 *eventually, but reaction too slow*

High temperature = low yield of SO_3 *obtained quickly*

In practice a compromise temperature of 500°C is used.

(ii) *Pressure*

The production of sulphur trioxide is also accompanied by a reduction in volume: $(2 + 1) = 3$ volumes of sulphur dioxide and oxygen produce only 2 volumes of sulphur trioxide. So

High pressure = high yield of SO_3 *obtained more quickly*

(c) The decomposition of dinitrogen tetraoxide
Brown nitrogen dioxide gas can be collected and trapped in a gas syringe. This gas is in equilibrium with pale-yellow dinitrogen tetraoxide gas according to the equation

$$2NO_2(g) \rightleftharpoons N_2O_4(g) + Heat$$
$$\text{dark brown} \qquad \text{pale yellow}$$

By apply Le Chatelier's principle, the reader should now be able to predict whether the equilibrium mixture would turn more pale (more N_2O_4) or darker (more NO_2)
  (i) on cooling the mixture down,
  (ii) on pushing hard on the plunger of the syringe to increase the pressure on the gas.

SUMMARY OF CHAPTER 6

The rate of reaction can be measured by following any convenient change that occurs during the reaction.

In a graph of that change plotted against time (time on the horizontal axis), the steeper the line the greater the rate.

Rate increases with increased temperature, with increased concentration, with increased surface area of the solid or in the presence of a catalyst.

These factors can be explained in terms of movement and collision of reacting particles.

Reactions which can go in either direction are called reversible reactions.

Such reactions eventually reach a state of dynamic equilibrium.

At equilibrium, rate of forward reaction = rate of backward reaction, and concentration of any chemical remains constant.

The effect of change of conditions on an equilibrium reaction can be predicted using Le Chatelier's principle.

CHAPTER 7

REDOX REACTIONS

In this chapter we shall see that most chemical reactions involve one chemical being 'reduced' and another 'oxidised'. Such reactions are termed 'reduction–oxidation' or *redox* reactions.

Our muscles tightening, food going bad, dynamite exploding, coal, wood or oil burning, colours bleaching, and a nail rusting all provide examples of redox reactions.

7.1 OXIDATION

If a strip of copper metal is held with tongs in a strong flame, a crumbly black skin forms on its surface and its weight increases. Surface copper has reacted with oxygen from the air to form copper(II) oxide, CuO, as a black ionic solid:

$$2Cu(s) + O_2(g) \longrightarrow 2CuO(s)$$

copper   oxygen   copper(II)
metal   oxide

Since copper has gained oxygen, we say that the copper has been '*oxidised*'. If, instead of air, copper is heated with sulphur, a similar black ionic compound results, this time of copper(II) sulphide, CuS:

$$Cu(s) + S(s) \longrightarrow CuS(s)$$

Because both reactions are similar, copper is again said to have been 'oxidised', even though no oxygen is involved in this latter reaction. Oxygen is therefore not essential in oxidation.

In each of the reactions below, for example, the metal is oxidised.

$$2Mg(s) + O_2(g) \longrightarrow 2MgO(s)$$
$$2Ag(s) + S(s) \longrightarrow Ag_2S(s)$$
$$2Al(s) + 3Cl_2(g) \longrightarrow 2AlCl_3(s)$$

$$Zn(s) \quad + \quad I_2(s) \longrightarrow ZnI_2(s)$$
$$3Mg(s) \quad + \quad N_2(g) \longrightarrow Mg_3N_2(s)$$
$$2FeCl_2(s) \quad + \quad Cl_2(g) \longrightarrow 2FeCl_3(s)$$

Oxidation probably seems a strange, if not an unfortunate, name to give to reactions which frequently do not involve oxygen. Over the years, the word 'oxidation' has been applied to a wider and wider range of reactions and, through use, the name has stuck.

Each of the above example involves the formation of ions. Oxygen, sulphur, chlorine, iodine and nitrogen have become stable negative ions by pulling electrons from the metals. The metals are said to have been oxidised because electrons have been removed from them, so

Oxidation is the removal of electrons

Removal of electrons from the metal (i.e. the oxidation of the metals) becomes more obvious if more detailed equations are written. For example, consider the following.

(i) $\qquad\qquad$ (electrons taken by oxygen)

$$2Mg(s) \quad + \quad O_2(g) \longrightarrow 2\,Mg^{2+}\,O^{2-}\,(s)$$

Electrons removed from magnesium atoms, therefore magnesium has been oxidised.

(ii) $\qquad\qquad$ (electrons taken by sulphur)

$$2Ag(s) \quad + \quad S(s) \longrightarrow Ag^{+}_2\,O^{2-}\,(s)$$

Electrons removed from silver atoms, silver has been oxidised.

(iii) $\qquad\qquad$ (electrons taken by iodine)

$$Zn(s) \quad + \quad I_2(s) \longrightarrow Zn^{2+}\,I^{-}_2\,(s)$$

Electrons removed from zinc atoms, zinc has been oxidised.

Elements such as oxygen, sulphur, nitrogen, chlorine and iodine are found to the right of the Periodic Table (see Section 2.6). All tend to become stable by gaining electrons to form negative ions—they are *electronegative.* Section 10.1(b) shows that this pull for electrons, or *electronegativity*, increases in the directions shown in Table 7.1. Thus a fluorine atom pulls electrons more strongly than does a chlorine atom, which itself pulls electrons more strongly than does an atom of sulphur.

Electronegativity shows up in covalent bonds. In carbon dioxide molecules, the shared electrons of the covalent bonds are partially pulled away from the carbon atom by the more electronegative oxygen atoms.

Table 7.1 *part of the Periodic Table, showing electronegativity change*

Group number	4	5	6	7

increase in ⟶
electronegativity

C	N	O	F	↑ increase
Si	P	S	Cl	in
			Br	electronegativity
			I	

Additionally, the electronegativity of hydrogen is similar to that of carbon.

O ＝ C ＝ O ⟵ ⟶   Shared electrons pulled towards oxygen atoms

Thus in the reaction

$$C(s) + O_2(g) \longrightarrow CO_2(g)$$

the oxygen atoms in the carbon dioxide product partially pull electrons from carbon, hence carbon has been oxidised.

Oxygen is appreciably more electronegative than hydrogen. In water molecules, electrons will be pulled slightly from hydrogen atoms.

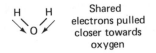

Shared electrons pulled closer towards oxygen

Thus hydrogen gas, when it burns in oxygen to form water, has been oxidised.

Similarly the reaction

$$H_2(g) + Cl_2(g) \longrightarrow 2HCl(g)$$

involves oxidation of hydrogen by more electronegative chlorine atoms.

H — Cl ⟶   Electrons pulled closer towards chlorine

7.2 REDUCTION

To *add* electrons is the very opposite of oxidation. The name given to this process is *reduction*:

Reduction is the addition of electrons

For example, consider the following.

(i)

(electrons gained by oxygen)

$$2Mg(s) + O_2(g) \longrightarrow 2\,Mg^{2+}\,O^{2-}\,(s)$$

Oxygen has been reduced.

(ii)

(electrons gained by sulphur)

$$2Ag(s) + S(s) \longrightarrow Ag^+{}_2\,S^{2-}\,(s)$$

Sulphur has been reduced.

(iii) $\quad H_2(g) + Cl_2(g) \longrightarrow 2\,H{-}Cl\,(g)$

(electrons now pulled closer to the more electronegative chlorine atom)

Chlorine has been reduced.

7.3 REDOX REACTIONS

In each of the above reactions, the atom *gaining* electrons was reduced. However, the gained electrons had to be removed from some other atom which must therefore have been oxidised. If something gains electrons (is reduced), something else must lose electrons (is oxidised). In all the reactions illustrated in this chapter, electrons are transferred *from* one atom (oxidised) *to* another (reduced). Such electron transfer reactions are termed reduction–oxidation or *redox* reactions.

7.4 OXIDISING AGENTS AND REDUCING AGENTS

Oxidation is the removal of electrons. A chemical which performs this removal is an *oxidising agent*, so

Oxidising agents are electron removers

Potassium iodide solution, KI(aq), is colourless. Iodine solution, I_2(aq), is brown. The following chemicals turn a clear solution of potassium iodide brown, indicating removal of electrons from the iodide ions to form brown iodine solution, according to the equation

$$2I^-(aq) - 2e \longrightarrow I_2(aq)$$

The chemicals are
potassium permanganate and dilute acid,
hydrogen peroxide and dilute acid,
potassium chlorate and dilute acid,

copper(II) sulphate, and
iron(III) chloride.

These chemicals, regardless of how they react or what they form, by *removing electrons* from I^- ions are behaving as oxidising agents.

Conversely, we can write

Reducing agents are electron donors

When sodium thiosulphate is added to a solution of iodine, the brown colour immediately clears. This suggests that the sodium thiosulphate has donated electrons to I_2 molecules to produce colourless I^- ions,

$$I_2(aq) + 2e \longrightarrow 2I^-(aq)$$

Sodium thiosulphate is therefore a reducing agent.

7.5 EXAMPLES

It is easy to confuse what has been reduced with reducing agent, and what has been oxidised with oxidising agent. The following questions help to decide which chemical is doing what in a redox reaction.

Which chemical has lost electrons? This chemical has been *oxidised.*
If it has lost electrons it must have
donated electrons to something. It is therefore a *reducing agent.*
Which chemical has gained electrons? This chemical has been *reduced.*
If it has gained electrons it must have
taken electrons from something. It is therefore an *oxidising agent.*

This is best illustrated by example.

(a) The violent reaction between sodium and chlorine

$$2Na(s) + Cl_2(g) \longrightarrow 2\,Na^+\,Cl^-\,(s)$$

Sodium has
    lost electrons ——— sodium has been oxidised.
    donated electrons ——— sodium is a reducing agent.

Chlorine has
    gained electrons ——— chlorine has been reduced.
    removed electrons ——— chlorine is an oxidising agent.

(b) The reaction of zinc with copper(II) chloride solution

If a strip of zinc metal is dipped into copper(II) chloride solution, a pink metallic copper colour appears on its surface. In this reaction, zinc dissolves to form ions as copper atoms are deposited.

$$Zn(s) + Cu^{2+}Cl^-_2\,(aq) \longrightarrow Cu(s) + Zn^{2+}Cl^-_2\,(aq)$$

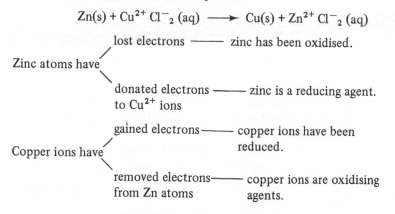

Zinc atoms have
- lost electrons —— zinc has been oxidised.
- donated electrons —— zinc is a reducing agent. to Cu^{2+} ions

Copper ions have
- gained electrons —— copper ions have been reduced.
- removed electrons from Zn atoms —— copper ions are oxidising agents.

(c) The reaction between phosphorus and chlorine

When dry chlorine gas is passed slowly over a piece of warm phosphorus in a strong glass tube, the phosphorus begins to burn gently to form a vapour. This condenses to a pale-yellow covalent liquid further down the tube. The liquid is phosphorus trichloride, PCl_3.

$$2P(s) + 3Cl_2\,(g) \longrightarrow 2PCl_3\,(l)$$

Chlorine atoms, being more electronegative than phosphorus, pull the shared electrons of this product from the central phosphorus atom.

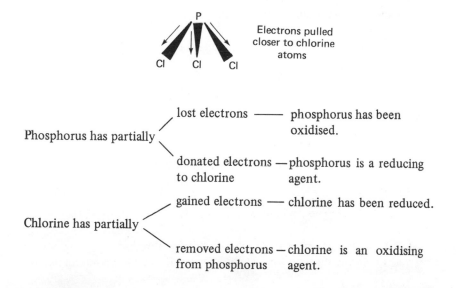

Electrons pulled closer to chlorine atoms

Phosphorus has partially
- lost electrons —— phosphorus has been oxidised.
- donated electrons to chlorine —— phosphorus is a reducing agent.

Chlorine has partially
- gained electrons —— chlorine has been reduced.
- removed electrons from phosphorus —— chlorine is an oxidising agent.

(d) The reaction between carbon and sulphur

When carbon and sulphur are heated together in a furnace, carbon disulphide, CS_2, is formed as a foul-smelling covalent liquid.

$$C(s) + 2S(s) \longrightarrow CS_2(l)$$

Shared electrons of the product are pulled away from carbon by the more electronegative sulphur atoms.

Carbon has partially
- lost electrons —— carbon has been oxidised.
- donated electrons — carbon is a reducing to sulphur agent.

Sulphur has partially
- gained electrons —— sulphur has been reduced.
- removed electrons — sulphur is an oxidising from carbon agent.

SUMMARY OF CHAPTER 7

Oxidation is the removal of electrons.

Reduction is the addition of electrons.

Whenever oxidation occurs, something must be simultaneously reduced; such reactions are called redox reactions.

An oxidising agent is an electron remover.

A reducing agent is an electron donor.

IONS AND ELECTRICITY

Any small electrically charged particle is called an *ion*. Section 3.3(f) showed how elements to the left of the Periodic Table, the metals, tend to form positive ions by losing electrons. Non-metals (found to the right of the Table) can form negative ions by electron gain.

8.1 IONIC FAVOURABILITY

For positive ions at least, it is possible to arrange an order of reactivity in which the metals that most readily ionise are at the top, with atoms unwilling to ionise at the bottom. A very approximate order of reactivity can be obtained by comparing the vigour of reaction of different metals with air and with water (see Table 8.1). A lump of potassium metal, for example, reacts explosively with water and would be placed towards the top of a reactivity list. Gold is totally unaffected even by boiling water or steam. It therefore appears at the bottom of this list.

When metals react, they *lose* electrons to form positive ions. Reactive elements towards the top of Table 8.1 are therefore those which easily give away electrons; they are the metals which most easily form ions. The less-reactive metals at the bottom more favourably remain as atoms and have the least tendency to ionise. It is helpful to visualise the reactivity series of Table 8.1 in terms of ionic or atomic *favourability*, as shown in Table 8.2.

8.2 OCCURRENCE OF THE METALS IN THE EARTH

The only metals in this list that are found *native*, that is that are found in the Earth as the free element, are copper, silver and gold. These are the elements which are more favourable as atoms. The others, being higher in the series and thus more favourable as ions, are found in nature only as ions. For example calcium in calcium carbonate (chalk, limestone, marble), $Ca^{2+}CO_3^{2-}$, potassium in the chloride K^+Cl^-, iron in the sulphide $Fe^{2+}S^{2-}$.

Table 8.1 *approximate order of reactivity for metals*

Element		Reaction with water	Reaction with air
Potassium	K	a violent reaction even with cold water as the metal rushes around the surface in a ball of flame	bursts into brilliant flame in warm air
Calcium	Ca	steadily gives off a stream of hydrogen bubbles	bursts into brilliant flame on heating in air
Magnesium	Mg	reacts exceptionally slowly with cold water, but quite violently with steam	bursts into brilliant flame on heating in air
Zinc	Zn	reacts only with steam	gives a bright flame on heating in air
Iron	Fe	an incomplete, reversible reaction with steam	reacts only on heating strongly in air
Lead	Pb	a very slight reaction detectable with steam	reacts only on heating strongly in air
Copper	Cu	does not react, even with steam	reacts only on heating strongly in air
Silver	Ag	does not react, even with steam	an incomplete, reversible reaction on heating in air
Gold	Au	does not react, even with steam	does not react

Table 8.2 *ionic and atomic favourability of metals*

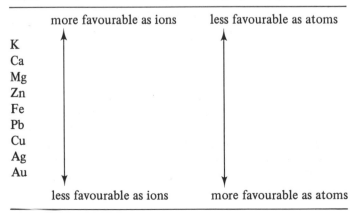

8.3 COMPETITION FOR OXYGEN

The experiments described in this section provide another method of determining this order of reactivity.

Two spatula measures of a powdered metal (comprising metal *atoms*) are mixed with the same quantity of the powdered oxide of a different metal (containing metal *ions*). The mixture is poured into an open crucible and a flame played briefly onto its surface in order to start any reaction.

A mixture of powdered zinc with copper(II) oxide, for example, quickly and violently erupts in brilliant flame.

$$Zn(s) + CuO(s) \longrightarrow Cu(s) + ZnO(s)$$

A reaction has occurred: zinc atoms took oxygen from the copper(II) oxide. The competition for oxygen was won by more reactive zinc atoms. Zinc is therefore rightly placed above copper in the reactivity series. In terms of atoms, ions and electrons, the zinc atoms have given away electrons to copper(II) ions:

(2 electrons
transferred)
$$Zn(s) + Cu^{2+}O^{2-}(s) \longrightarrow Cu(s) + Zn^{2+}O^{2-}(s)$$
atoms   ions   atoms   ions

Hence, we can write that
zinc atoms have lost electrons to become ions,
zinc is more favourable as ions;
copper ions have gained electrons to become atoms,
copper is more favourable as atoms.

In a similar experiment, a mixture of copper metal and zinc(II) oxide shows no sign of reaction

$$Cu(s) + ZnO(s) \longrightarrow \text{no reaction}$$

because zinc, the more favourable as ions, already exists as ions.

The results of similar experiments are summarised in Table 8.3.

8.4 ATOM-ION EXCHANGES IN SOLUTION

In a more peaceful and often beautiful experiment, a strip of one metal is dipped into a solution containing ions of another. For example, a strip of zinc metal (containing zinc atoms) is dipped into a solution of copper(II) sulphate solution (containing Cu^{2+} ions). Within a few minutes, a coating of pink new copper metal has formed on the surface of the zinc. Copper(II) ions are turning into copper atoms, presumably at the expense of the zinc atoms:

94

Table 8.3 *summary of various 'competition for oxygen' experiments*

Mixture	Reaction observed? (was heat + light given out?)	Conclusions
Mg + CuO	yes	more favourable as ions than Mg ⟍ ⟋ Cu should be placed above
Pb + Zno	no	less favourable as ions than Pb ⟋ ⟍ Zn should be placed below
Zn + PbO	yes	more favourable as ions than Zn ⟋ ⟍ Pb should be placed above
Fe + CuO	yes	more favourable as ions than Fe ⟋ ⟍ Cu should be placed above
Cu + PbO	no	less favourable as ions than Cu ⟋ ⟍ Pb should be placed below

$$\underset{\text{atoms}}{Zn(s)} + \underset{\text{ions}}{Cu^{2+}SO_4^{2-}(aq)} \xrightarrow{\text{(2 electrons transferred)}} \underset{\text{atoms ions}}{Cu(s) + Zn^{2+}SO_4^{2-}(aq)}$$

The equation can be simplified by omitting the SO_4^{2-} ions, which take no part in the reaction (they are the same at the start and finish). These are sometimes described as 'spectator' ions. So we can write

$$\underset{\text{atoms}}{Zn(s)} + \underset{\text{ions}}{Cu^{2+}(aq)} \xrightarrow{\text{(2 electrons transferred)}} \underset{\text{atoms ions}}{Cu(s) + Zn^{2+}(aq)}$$

Again, in this experiment, zinc is shown to be more favourable as ions and copper more favourable as atoms. The placing of zinc above copper in the reactivity series is justified. Similar experiments, outlined in Table 8.4, verify the positions of other metals.

All these reactions involve *electron transfer*. They are therefore all *redox* reactions (see Chapter 7).

Table 8.4 *summary of various atom-ion displacement experiments*

Metal strip (atoms)	Metal ions in solution*	Observation	Conclusions
Zn	Pb^{2+}	sparkling black crystals of lead form on the zinc strip	Zn — is more favourable as ions than — Pb / should be placed above
Mg	Zn^{2+}	shiny grey crystals of zinc form on the magnesium strip	Mg — is more favourable as ions than — Zn / should be placed above
Ag	Cu^{2+}	no change observed	Ag — is less favourable as ions than — Cu / should be placed below
Cu	Ag^+	sparkling crystals of silver form on the copper strip	Cu — is more favourable as ions than — Ag / should be placed above

*For example, the metal nitrate solution could be used.

8.5 THE ELECTROCHEMICAL SERIES

The experiments described in Section 8.4 involve ions in aqueous solution. The series of 'ionic favourability' in which the ions are in *aqueous solution* is called the *electrochemical series*. The more complete electrochemical series shown in Table 8.5 will be used throughout the remainder of this text.

8.6 ELECTRICAL CURRENT

The *circuit* shown in Fig. 8.1 converts electrical energy into light and heat energy. Electrons are simultaneously pushed out from the negative terminal of the cell and pulled in by the positive terminal (Fig. 8.2). The connecting metal wire finds electrons forced into one end and pulled out at the other. In this way, delocalised electrons are moved along the wire.

Table 8.5 *the electrochemical series for some common metals*

		more favourable as ions	less favourable as atoms
Lithium	Li	↑	↑
Potassium	K		
Calcium	Ca		
Sodium	Na		
Magnesium	Mg		
Aluminium	Al		
Zinc	Zn		
Iron	Fe		
Nickel	Ni		
Tin	Sn		
Lead	Pb		
Copper	Cu		
Mercury	Hg		
Silver	Ag		
Gold	Au	↓	↓
		less favourable as ions	more favourable as atoms

fig 8.1 *an electric circuit*

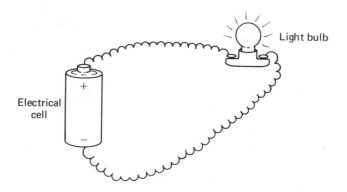

The enforced drift of electrons along the wire is an *electric current*.
It is often helpful to view the electrical supply as an electron pump

Pumping electrons out at the negative terminal

and

Sucking electrons in at the positive terminal

The greater the voltage of the supply the greater the force with which

fig 8.2 *a way of visualising electric current*

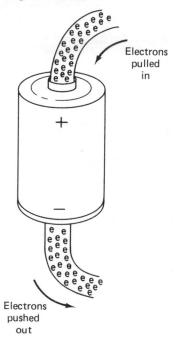

Electrons
pulled
in

Electrons
pushed
out

electrons are pumped out and sucked in. Diagrammatically an electric cell is represented as

where the longer line is the positive terminal.

8.7 THE CONDUCTION OF ELECTRICITY THROUGH CHEMICALS

(a) Conduction through elements
The circuit shown in Fig. 8.3 will test the electrical conductivity of a substance, simply by touching x and y onto an object and watching for a glow in the light. In this way we would find that metals (and graphite) do conduct electricity. Section 3.6 rationalised this in terms of the freely moving delocalised electrons within these substances. This experiment would also show the non-metals to be non-conductors.

(b) Conduction through compounds
Since they do not contain separate particles with a net charge, covalent

98

fig **8.3** *testing electrical conductivity*

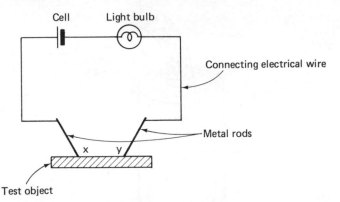

Test object

compounds are non-conductors. Ionically bonded compounds do contain charged ions. Section 3.9 showed that once these ions have been freed from their fixed positions in the solid, either by melting or by dissolving in water, ionic compounds will conduct.

Apparatus to test for electrical conductivity of ionic compounds is shown in Fig. 8.4. In this experiment the conducting rods (often composed of graphite) which dip into the molten compound are called *electrodes*. The electrode connected to the positive terminal of the cell is called the *anode* and that connected to the negative terminal the *cathode*.

fig **8.4** *the electrical conductivity of ionic compounds*

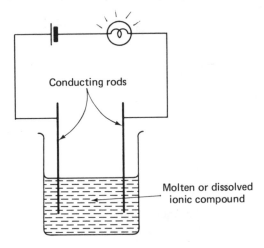

The experiments in sections (a) and (b) above show two quite different ways in which electricity can be conducted. Apart from getting a little warmer, the rod of metal or graphite is unaffected by the passage of electricity. However, when electricity is passed through molten lead bromide, for example, a considerable change is observed (Fig. 8.5).

fig 8.5 *conducting electricity through molten lead bromide*

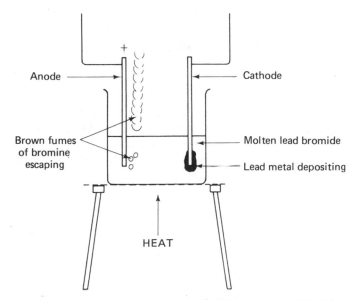

Around the cathode (negative electrode) shiny globules of lead begin to appear, whilst brown fumes of bromine pour from the anode. Similarly, when electricity is passed through molten sodium chloride, sodium metal and chlorine gas appear at the electrodes, whilst potassium and iodine separate out when electricity is forced through molten potassium iodide.

This decomposition of ionic compounds by the action of electricity is called *electrolysis*; the compounds have been *electrolysed* and are termed *electrolytes*. The apparatus in which the electrolysis occurs is called a *voltameter*. Substances can therefore be divided up according to how they conduct electricity as shown in Table 8.6.

8.8 EXPLAINING ELECTROLYSIS

The theory of ionic bonding given in Section 3.3 can be used to give a picture of what happens during electrolysis.

(a) The electrolysis of sodium chloride
Section 3.3 pictured sodium chloride as an array of positive sodium ions

Table 8.6 *electrical conduction through substances*

CONDUCTORS		NON-CONDUCTORS
Metals and graphite	Ionic compounds	Covalent compounds
(these conduct without decomposing)	(these are electrolytes, conducting only when the ions are free to move, decomposing as electricity flows)	(these do not conduct electricity)

(formed by sodium atoms losing electrons) attracted to negative chloride ions (formed by chlorine atoms gaining electrons). When sodium chloride is melted these ions are free to move.

Positive Na^+ ions will be attracted towards the cathode (negative) and negative Cl^- ions towards the anode (positive). In the electrical circuit, electrons are pushed out from the negative terminal of the cell onto the cathode. Electrons on the cathode will thus be pushed onto Na^+ ions, converting them into neutral Na atoms:

$$Na^+ + 1e \longrightarrow Na$$
$$\text{ions} \qquad\qquad \text{atoms}$$

Thus sodium metal collects at the cathode.

The positive terminal of the cell is sucking electrons in from the anode. Some electrons may be pulled by the anode from Cl^- ions, converting them into Cl atoms:

$$Cl^- \longrightarrow 1e + Cl$$
$$\text{ion} \qquad \text{taken} \quad \text{atom}$$
$$\qquad\quad \text{by}$$
$$\qquad\quad \text{anode}$$

Of course Cl atoms immediately combine to form Cl_2 molecules (see Section 3.4(b)), and the equation is more precisely written

$$2Cl^- \longrightarrow 2e + Cl_2$$

Thus chlorine gas bubbles from the anode.

Since positive ions are always attracted to the cathode, they are called *cations*. Negative ions, being attracted towards the anode, are called *anions*. In this example, sodium ions are cations and chloride ions are anions.

The electrolysis of sodium chloride may be summarised as follows.

At the cathode	*At the anode*
Na^+ ions attracted. These have electrons pushed on:	Cl^- ions attracted. These have electrons pulled off:
$$Na^+ + 1e \rightarrow Na$$	$$2Cl^- \rightarrow 2e + Cl_2$$
Sodium metal deposits.	Chlorine gas bubbles off.

(b) The electrolysis of molten lead bromide

This may be summarised as follows.

At the cathode	*At the anode*
Pb^{2+} ions attracted. These have electrons pushed on:	Br^- ions attracted. These have electrons pulled off:
$$Pb^{2+} + 2e \rightarrow Pb$$	$$2Br^- \rightarrow 2e + Br_2$$
Lead metal deposited.	Bromine formed.

(c) The electrolysis of molten aluminium oxide

Similarly, this may be summarised.

At the cathode	*At the anode*
Al^{3+} ions attracted. These have electrons pushed on:	O^{2-} ions attracted. These have electrons pulled off:
$$Al^{3+} + 3e \rightarrow Al$$	$$2O^{2-} \rightarrow 4e + O_2$$
Aluminium metal deposited.	Oxygen gas bubbles off.

8.9 THE COMPLICATION OF WATER

We have seen that ionic compounds will conduct electricity only once their ions are free to move. As an alternative to melting the compound, freely moving ions may also be obtained by dissolving in water. However, now the electrolysis is complicated by the presence of the water.

Water is an exceptionally feeble conductor of electricity. Only at high voltages does it conduct appreciably, decomposing to give hydrogen gas at the cathode and oxygen at the anode. It is surprising that water conducts electricity at all when it is supposed to be composed of neutral H_2O molecules (see Section 3.4(e)). It seems that water molecules can split into hydrogen ions, H^+, and hydroxide ions, OH^-, in a reversible reaction:

$$H_2O(l) \rightleftharpoons H^+(aq) + OH^-(aq)$$

However, the resulting equilibrium is heavily biased towards H_2O molecules. Experiments suggest that, on average, out of every ten million water molecules there is just one that has split into ions. Nevertheless, a beaker of water, holding billions of water molecules, will contain a vast number of hydrogen and hydroxide ions, and the presence of these ions has a marked effect on the electrolysis products. This is best illustrated by example.

(a) The electrolysis of aqueous sodium chloride

When a solution of common salt in water conducts electricity, there are four different ions in the liquid:

$$Na^+ \qquad Cl^- \qquad \text{in sodium chloride}$$

$$H^+(aq) \qquad OH^- \qquad \text{in water}$$

(i) At the cathode

Both Na^+ and H^+ ions will be attracted to the cathode. A glance at the electrochemical series (often abbreviated ECS) (see Section 8.5) shows that sodium is the more favourably ionic. Thus the sodium ions *remain* in solution *as ions*. Electrons are therefore pushed from the cathode onto hydrogen ions

$$2H^+(aq) + 2e \longrightarrow H_2(g)$$

converting them into hydrogen molecules. Hydrogen gas is seen to bubble from the cathode.

Whenever there is a choice of positive ions in an electrolysis, whichever is less favourable as ions is the one that is deposited at the cathode. Thus

The element which is lower in the electrochemical series is preferentially deposited

(iii) At the anode

At the anode, a choking pale-green gas is given off. This is chlorine. However, if the solution of sodium chloride is very dilute, experiment finds the anode gas to be oxygen. The anode will attract both Cl^- and OH^- ions. Electrons are pulled preferentially from Cl^- ions when these are in high concentration:

$$2Cl^-(aq) \longrightarrow 2e + Cl_2(g)$$

If chloride ions are not in high concentration, then electrons are taken from hydroxide ions to form oxygen gas at the anode:

$$4OH^-(aq) \longrightarrow 4e + 2H_2O(l) + O_2(g)$$

As a working rule,

Electrons are removed from hydroxide ions in preference to all other anions except chloride, bromide or iodide when these are present in concentrated solution

(iii) *Summary*

The electrolysis of aqueous sodium chloride can be summarised as follows.

(1) Concentrated aqueous sodium chloride

At the cathode
Na^+ and H^+ ions attracted. H^+ ions (lower in ECS) gain electrons:

$$2H^+ + 2e \rightarrow H_2$$

Hydrogen gas bubbles off. Na^+ ions left in solution.

At the anode
Cl^- and OH^- ions attracted. Cl^- ions (concentrated solution) give up electrons:

$$2Cl^- \rightarrow 2e + Cl_2$$

Chlorine gas bubbles off. OH^- ions left in solution.

This reaction is used in the industrial preparation of hydrogen gas, chlorine gas and sodium hydroxide solution.

(2) Dilute aqueous sodium chloride

At the cathode
Na^+ and H^+ ions attracted. H^+ ions (lower in ECS) gain electrons:

$$2H^+ + 2e \rightarrow H_2$$

Hydrogen gas bubbles off.

At the anode
Cl^- and OH^- ions attracted. OH^- ions (dilute solution) give up electrons:

$$4OH^- \rightarrow 4e + 2H_2O + O_2$$

Oxygen gas bubbles off.

(b) The electrolysis of concentrated aqueous potassium iodide
There are four different ions:

$$K^+ \qquad I^- \qquad \text{in potassium iodide}$$

$$H^+(aq) \qquad OH^- \qquad \text{in water}$$

At the cathode
K^+ and H^+ ions attracted. H^+ ions (lower in ECS) gain electrons:

$$2H^+ + 2e \rightarrow H_2$$

Hydrogen gas bubbles off.

At the anode
I^- and OH^- ions attracted. I^- ions (concentrated solution) give up electrons:

$$2I^- \rightarrow 2e + I_2$$

A brown solution of iodine collects around the anode.

(c) The electrolysis of dilute sulphuric acid
There are three different ions:

$$2H^+(aq) \quad SO_4^{2-} \quad \text{in sulphuric acid}$$

$$H^+(aq) \quad OH^- \quad \text{in water}$$

At the cathode
H^+ ions are the only cations present. Therefore

$$2H^+ + 2e \rightarrow H_2$$

Hydrogen gas bubbles off.

At the anode
SO_4^{2-} and OH^- ions attracted. OH^- ions preferentially give up electrons:

$$4OH^- \rightarrow 4e + 2H_2O + O_2$$

Oxygen gas bubbles off.

(d) The electrolysis of copper(II) sulphate solution

There are four different ions:

$$Cu^{2+} \quad SO_4^{2-} \quad \text{in copper sulphate}$$

$$H^+(aq) \quad OH^- \quad \text{in water}$$

At the cathode
Cu^{2+} and H^+ ions attracted. Cu^{2+} ions (lower in ECS) gain electrons:

$$Cu^{2+} + 2e \rightarrow Cu$$

Pink metallic copper collects on the cathode.

At the anode
SO_4^{2-} and OH^- ions attracted. OH^- ions preferentially give up electrons:

$$4OH^- \rightarrow 4e + 2H_2O + O_2$$

Oxygen gas bubbles off.

8.10 INFLUENCE OF THE ELECTRODES

In the examples of electrolysis met so far, the electrodes have done no more than allow electrons to pass through. Platinum makes an excellent inert electrode of this type. Strips and rods of many cheaper materials are often used as electrodes, but sometimes these can alter the electrolysis. Examples are given below.

(a) The electrolysis of molten aluminium oxide using graphite electrodes

At the cathode
Molten aluminium collects at this electrode as expected.

At the anode
O^{2-} ions attracted, which have electrons pulled off:

$$2O^{2-} \rightarrow 2e + O_2$$

However, this oxygen immediately reacts with the graphite (carbon) anode to form bubbles of carbon dioxide, and the anode is gradually eaten away.

This electrolysis is the basis of the industrial manufacture of aluminium (see Section 11.8).

(b) The electrolysis of aqueous copper(II) sulphate using copper electrodes

At the cathode
Copper metal is deposited as expected:

At the anode
SO_4^{2-} and OH^- ions are attracted. With a copper anode, electrons are pulled, not from SO_4^{2-} nor from OH^- ions, but from copper atoms of the anode itself:

$$Cu^{2+} + 2e \rightarrow Cu$$

$$Cu \rightarrow 2e + Cu^{2+}$$

The anode therefore dissolves to form aqueous Cu^{2+} ions.

Thus the copper anode dissolves into the solution as fast as the copper cathode grows, leaving the concentration of the solution constant. This reaction is used industrially to obtain very pure copper, called *electrolytic copper*. Starting with a thin pure copper cathode and using a lump of impure copper as the anode, only the pure metal transfers to the cathode in the electrolysis. Thus a pure copper cathode is obtained from an impure copper anode by this technique of *electrolytic refining*.

(c) The electrolysis of sodium chloride solution using a mercury cathode
Mercury exists as a molten metal even at room temperature.

At the mercury cathode
Na^+ and H^+ ions are attracted. H^+ ions are lower in the ECS, and would be expected to gain electrons. However, sodium metal can dissolve in mercury to form *sodium amalgam* (a solution of any metal in mercury is called an amalgam). With mercury as the cathode, it is Na^+ ions which gain electrons:

At the anode
Chlorine gas bubbles from the anode as expected.

$$Na^+ + 1e \rightarrow Na$$

Sodium atoms thus formed immediately dissolve in the mercury to
form sodium amalgam.

This electrolysis is the basis of the industrial manufacture of sodium
hydroxide (see Section 11.8(c)).

8.11 ELECTROPLATING

The technique of electroplating enables metals such as iron to be coated
with a thin protective layer of a more expensive corrosion-resistant metal
such as silver. Common examples are spoons and bowls labelled 'EPNS'
(electroplated nickel silver), chromium plating on the steel bumpers of
cars, gold-plated jewellery. The plating is generally exceptionally thin—
often no more than one thousand atoms thick.

For example, a brass object can be plated with a silvery coating of
nickel by making it the cathode in the circuit shown in Fig. 8.6. It is essential that the surface of the object is clean and grease free, and this is

fig 8.6 *nickel plating*

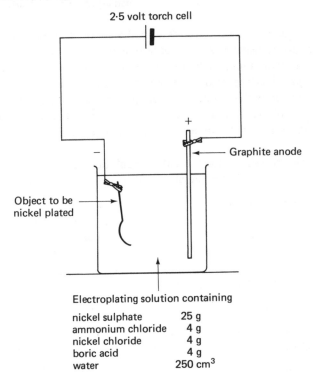

2·5 volt torch cell

+

− ────── Graphite anode

Object to be ───→
nickel plated

Electroplating solution containing

nickel sulphate	25 g
ammonium chloride	4 g
nickel chloride	4 g
boric acid	4 g
water	250 cm^3

achieved by washing in detergent, rinsing in pure water, drying and rubbing with fine sandpaper before setting up the electrolysis. The extra ingredients in the plating solution serve to control acidity.

8.12 ELECTRICITY FROM CHEMICAL REACTIONS

Many chemical reactions give out heat energy. The experiment in Fig. 8.7 shows how this energy may be obtained alternatively in the form of electricity.

fig 8.7 *electricity from a chemical reaction*

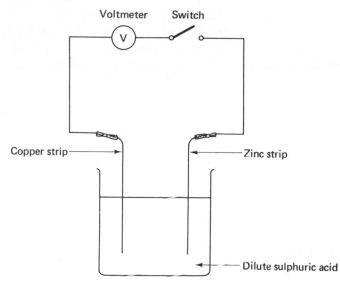

When the switch is in the 'off' position, the zinc rod steadily gives off bubbles of hydrogen as it reacts with the dilute acid. The copper rod shows no sign of reaction. When the switch is 'on', bubbles of hydrogen now begin to pour from the copper electrode with fewer appearing from the zinc rod, the zinc rod continues to dissolve and an electrical voltage registers on the meter. This is a simple electrical *cell*, discovered by Volta nearly two hundred years ago.

Replacing the zinc rod with other metals produces different voltages (see Table 8.7). It seems that

The greater the separation of the two metals in the electrochemical series the greater the voltage

In practice, electricity is never produced by such an inefficient method. In the above example, the zinc rod dissolves slowly even when no current

Table 8.7 *typical voltages obtained from experiments with simple cells*

Electrodes*	Voltage reading/volts
Copper with magnesium	1.2
Copper with zinc	0.8
Copper with iron	0.5
Copper with lead	0.1

*In each case, copper is found to be the positive terminal.

is being produced. Such a battery would go 'flat' within ten minutes of it leaving the factory. Other slightly more complicated cells overcome this problem, as well as giving an insight into what might be happening amongst the atoms and ions concerned in order to produce these electric currents.

(a) The Daniell cell

The original Daniell cell is still to be seen at King's College, London, where Daniell was professor of chemistry between 1831 and 1845. Such a cell is shown diagrammatically in Fig. 8.8.

Zinc is more favourable as ions than copper (it is higher in the electrochemical series). Atoms in the zinc rod tend to lose electrons, leaving zinc ions to dissolve into the solution:

$$Zn(s) \longrightarrow 2e + Zn^{2+}(aq)$$

fig 8.8 *the Daniell cell*

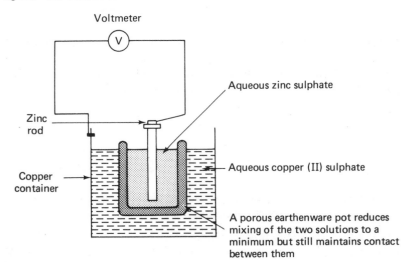

Voltmeter

Zinc rod

Copper container

Aqueous zinc sulphate

Aqueous copper (II) sulphate

A porous earthenware pot reduces mixing of the two solutions to a minimum but still maintains contact between them

This means that electrons are building up on the zinc terminal. Copper is more favourable as atoms than zinc. Copper(II) ions tend to pull electrons from the metal container to form copper atoms, which deposit on the inner surface of the container:

$$Cu^{2+}(aq) + 2e \longrightarrow Cu(s)$$

This means that there is a deficiency of electrons in the copper of the container. By connecting zinc (with a build-up of electrons) to copper (with a deficiency of electrons) by a conducting wire, electrons will flow from zinc to copper. Thus an electric current flows.

(b) The dry cell
A torch or radio battery is frequently a *dry cell*. A gelatinous paste of ammonium chloride and flour is used in place of the two solutions of the Daniell cell, and a rod of graphite replaces the copper.

8.13 SACRIFICIAL ZINC

Although it has the advantage of being a strong and comparatively cheap metal, iron corrodes very easily by rusting (see Section 11.6(d)) turning into iron(III) ions. Zinc, being above iron in the electrochemical series (see Section 8.5), ionises in preference to iron. For this reason, zinc blocks are frequently bolted on to the iron keels of ships so that the two metals are in electrical contact. In preference to the reaction

$$Fe(s) \longrightarrow Fe^{3+}(aq) + 3e$$

the electrons will now be lost from zinc

$$Zn(s) \longrightarrow Zn^{2+}(aq) + 2e$$

leaving the iron keel uncorroded. It is then a relatively simple matter to replace these *sacrificial* zinc blocks.

Galvanised iron, used for metal water tanks, electricity pylons and motorway crash barriers, consists of steel coated with a layer of zinc for this same purpose.

SUMMARY OF CHAPTER 8

Metals high in the electrochemical series are favourable as ions; those low in the series are favourable as atoms.

When electricity is passed through a molten or dissolved ionic compound, electrolysis occurs. Such a compound is an electrolyte.

Electrolysis is always accompanied by decomposition at the electrodes. Which elements deposit on the electrodes depend upon the relative positions of the cations in the electrochemical series, upon the concentration of the solution and upon the nature of the electrodes.

Electricity may be obtained from chemical reactions between two different conductors in contact with electrolyte.

CHAPTER 9

ACIDS

To many people, chemistry is to do with acids, and acids are dangerous corrosive liquids.

In fact, acids can be gases, liquids or solids; some are exceptionally corrosive but some we eat and drink, some are the essential units from which our bodies are made. In most swimming baths, we even bathe in acid.

9.1 WHAT ACIDS DO

A great many substances, both natural and man-made, behave in a similar way. These substances all
  (i) taste sour,
 (ii) react with zinc and other metals high in the electrochemical series to give hydrogen gas,
(iii) change the colour of many natural and synthetic dyes,
(iv) generally react with calcium carbonate to produce carbon dioxide, and
 (v) find these properties destroyed by reaction with metal oxides or metal hydroxides.

(i) *Taste*

Lemon juice, an acid-drop sweet, cream of tartar or vinegar all taste sour.

(ii) *Reaction with zinc and other metals high in the electrochemical series*

With such metals, acids produce hydrogen gas. The apparatus used is shown in Fig. 9.1. Because of the explosion risk with hydrogen gas, flames must be kept well away from the flask. When a flame is held to the mouth of the test tube some distance from the apparatus, the gas explodes with a 'pop'. This is a characteristic of hydrogen, and is used as a test for this gas.

fig 9.1 *reaction of zinc with acid to produce hydrogen*

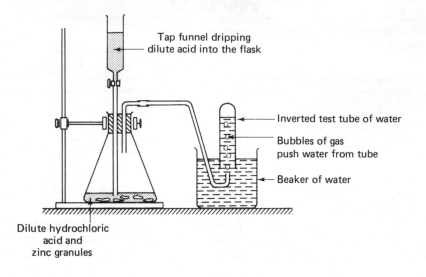

Tap funnel dripping dilute acid into the flask

Inverted test tube of water

Bubbles of gas push water from tube

Beaker of water

Dilute hydrochloric acid and zinc granules

(iii) *Effect on the colour of many dyes*

Dilute sulphuric acid, dilute hydrochloric acid, vinegar and many other substances turn the orange pattern on kitchen towels grey, the colour of purple cabbage red, the purple of litmus root red, and the synthetic dye phenolphthalein from red to colourless.

(iv) *Reaction with calcium carbonate*

With calcium carbonate, acids generally produce carbon dioxide. Using the apparatus shown in Fig. 9.1, dilute hydrochloric acid may be dripped onto small lumps of marble (marble is one form of calcium carbonate). When a few drops of clear lime water are added to the resulting tube of gas, a milkiness appears. This is a characteristic of carbon dioxide frequently used as a test for this gas.

(v) *Destruction of these properties*

These properties (i) to (iv) mentioned are destroyed (neutralised) if the acid reacts with a metal oxide or a metal hydroxide (see Fig. 9.2).

Many substances show these five properties. They are classed together as *acids*. Table 9.1a lists just a few of the many substances which show acidic properties, and Table 9.1b give the formulae for some acids.

fig 9.2 *neutralising acidity*

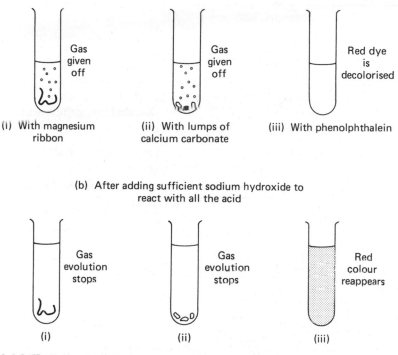

(a) Test tubes of dilute hydrochloric acid

(i) With magnesium ribbon — Gas given off

(ii) With lumps of calcium carbonate — Gas given off

(iii) With phenolphthalein — Red dye is decolorised

(b) After adding sufficient sodium hydroxide to react with all the acid

(i) — Gas evolution stops

(ii) — Gas evolution stops

(iii) — Red colour reappears

9.2 WHAT IS COMMON TO ACIDS?

Since acids have several properties in common, it seems reasonable that they might contain something in common. The formulae in Table 9.1b suggest that hydrogen might be the important ingredient. On the other hand, methanol CH_4O, methane CH_4, toluene C_7H_8 and glucose $C_6H_{12}O_6$ each contain hydrogen but *do not* show acidic properties. However, these substances do not conduct electricity. Solutions of the acids do conduct, with hydrogen gas bubbling from the cathode in every case.

At the cathode

$$2H^+(aq) + 2e \longrightarrow H_2(g)$$

This suggests that all acids contain the hydrogen ion, $H^+(aq)$. This ion is responsible for acidic properties.

9.3 THE NEED FOR WATER

Dry hydrogen chloride gas (HCl) dissolved in dry liquid toluene does not show acidic properties. Such a solution

Table 9.1

(a) *substances showing acidic properties*

	Sulphuric acid	
	Hydrochloric acid	
	Nitric acid	
	Lemon juice	contains citric acid
	Coca-Cola	contains phosphoric(V) acid
	Orange juice	contains citric acid
	Digestive juices	contain hydrochloric acid
	Vinegar	contains ethanoic acid
Aqueous solutions of:	Carbon dioxide	forms carbonic acid
	Chlorine gas	forms hydrochloric acid
	Phosphorus(V) oxide	forms phosphoric(V) acid

(b) *formulae of these acids*

Sulphuric acid	H_2SO_4
Hydrochloric acid	HCl
Nitric acid	HNO_3
Citric acid	$C_6H_8O_7$
Phosphoric(V) acid	H_3PO_4
Ethanoic acid	$C_2H_4O_2$
Carbonic acid	H_2CO_3

(i) when added to paper impregnated with blue litmus dye does not turn the paper red,

(ii) when added to pieces of zinc metal does not produce bubbles of hydrogen gas;

(iii) when added to pieces of marble does not produce bubbles of carbon dioxide gas.

However, stirring the contents of each tube with a few drops of water immediately produces the expected acid reaction. Pure dry ethanoic acid behaves in just the same way: no acidity is shown until water is added. Furthermore, neither dry hydrogen chloride in dry toluene, nor dry ethanoic acid conduct electricity. They do not contain ions even though they do contain hydrogen.

The experiments above suggest that hydrogen ions, and thus acidity, can only occur in the presence of water. In fact experiments now indicate that, in the absence of water, H^+ ions are too reactive to exist. The H^+ ion, being a hydrogen atom that has lost its electron, is simply one isolated proton (Fig. 9.3).

H^+ is the smallest possible positive ion, one hundred thousand times

fig 9.3 *loss of an electron from the hydrogen atom*

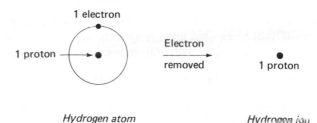

Hydrogen atom   *Hydrogen ion*

smaller than a sodium ion. Such a small charged particle would be attracted very strongly towards the electrons of anything. For example, the reaction

$$H-Cl(g) \longrightarrow H^+ + Cl^-$$

cannot occur, because the highly concentrated positive charge of H^+ would immediately join back with the electrons of Cl^- to reform covalent H–Cl.

With water, H^+ ions are pulled into the H_2O molecules to form much larger, stable *hydroxonium ions*, H_3O^+:

$$H^+ + H_2O \longrightarrow H_3O^+$$

Hydrogen ions cannot exist on their own, but they can exist combined with water. Thus hydrogen ions must always be shown as $H^+(aq)$ *or* as H_3O^+.

When hydrogen chloride gas meets water, heat is given out. A thermometer with one drop of water held on its bulb, when lowered into a flask of hydrogen chloride gas, records a rise in temperature of about 20°C.

On finding water, the hydrogen chloride is able to ionise, and the heat given out is the heat of this reaction:

$$HCl(g) + H_2O(l) \longrightarrow H_3O^+(aq) + Cl^-(aq)$$

Similarly, water allows other acids to ionise:

$$HNO_3(l) + H_2O(l) \longrightarrow H_3O^+(aq) + NO_3^-(aq)$$
nitric
acid

$$H_2SO_4(l) + 2H_2O(l) \longrightarrow 2H_3O^+(aq) + SO_4^{2-}(aq)$$
sulphuric
acid

In the case of mixing either concentrated nitric or concentrated sulphuric acids with water, so much heat is given out that it is important always to stir acid into water, not water into acid, so that any splashing or cracking caused by extreme localised heating would not involve the container of corrosive pure acid.

A quick glance through chemistry books and scientific papers shows the symbol $H^+(aq)$ to be used more frequently than H_3O^+. Nevertheless, it is a good idea to be familiar with both forms:

$$HCl(g) \quad + Aq \longrightarrow \quad H^+(aq) \quad + Cl^-(aq)$$

or

$$HCl(g) \quad + H_2O(l) \longrightarrow \quad H_3O^+(aq) + \quad Cl^-(aq)$$

$$H_2SO_4(l) + Aq \longrightarrow \quad 2H^+(aq) \quad + SO_4{}^{2-}(aq)$$

or

$$H_2SO_4(l) + 2H_2O(l) \longrightarrow 2H_3O^+(aq) + SO_4{}^{2-}(aq)$$

In the remainder of this book, the symbol $H^+(aq)$ will be used.

9.4 REACTIONS OF THE HYDROGEN ION

Having now decided that the presence of $H^+(aq)$ is the requirement for acidity, it should be possible to write equations representing the reaction of $H^+(aq)$.

(a) Reaction with metals
Zinc dissolves in acids as hydrogen gas bubbles from it:

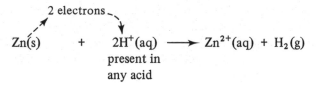

$$Zn(s) \quad + \quad 2H^+(aq) \longrightarrow Zn^{2+}(aq) + H_2(g)$$
present in
any acid

Thus zinc is more favourable as ions than hydrogen. Hydrogen should be placed below zinc in the electrochemical series (see Section 8.4). Experiments find that lead is the lowest metal in the electrochemical series to liberate hydrogen gas from an acid:

$$Pb(s) + \quad 2H^+(aq) \longrightarrow Pb^{2+}(aq) + H_2(g)$$
present in
any acid

Thus lead is *just* more favourably ionic than hydrogen, and therefore hydrogen is placed immediately below lead in the electrochemical series.

All metals above hydrogen in the electrochemical series will displace hydrogen from dilute aqueous acids. Examples using specific acids and metals are given below.

(i) *Magnesium with dilute sulphuric acid*

$$Mg(s) + H_2SO_4(aq) \longrightarrow MgSO_4(aq) + H_2(g)$$

magnesium
sulphate

This can be written more simply as

$$Mg(s) + 2H^+(aq) \longrightarrow Mg^{2+}(aq) + H_2(g)$$

(ii) *Calcium with dilute hydrochloric acid*

$$Ca(s) + 2HCl(aq) \longrightarrow CaCl_2(aq) + H_2(g)$$

calcium
chloride

This can be written more simply as

$$Ca(s) + 2H^+(aq) \longrightarrow Ca^{2+}(aq) + H_2(g)$$

It can be seen from these examples that the anions (negative ions) take no part in the reaction. Chloride ions in example (ii) are present as Cl^- at the start and finish, for example, and might just as well be excluded from the equation.

(b) Reaction with carbonates

Acids react with the carbonate ion, CO_3^{2-}, to produce carbon dioxide.

(i) *Calcium carbonate with dilute hydrochloric acid*

$$CaCO_3(s) + 2HCl(aq) \longrightarrow CaCl_2(aq) + H_2O(l) + CO_2(g)$$

calcium
chloride

This can be written more simply as

$$CO_3^{2-}(aq) + 2H^+(aq) \longrightarrow H_2O(l) + CO_2(g)$$

(Ca^{2+} and Cl^- ions are present at the start and finish of the reaction, so might just as well be excluded from the equation).

(ii) *Sodium carbonate solution with dilute sulphuric acid*

$$Na_2CO_3(aq) + H_2SO_4(aq) \longrightarrow Na_2SO_4(aq) + H_2O(l) + CO_2(g)$$

sodium
sulphate

This can be written more simply

$$CO_3^{2-}(aq) + 2H^+(aq) \longrightarrow H_2O(l) + CO_2(g)$$

(c) Reactions with indicators
Dyes giving specific colours when acid is added (see Section 9.1) are called indicators. They can be used as a quick test for the presence of acid. Phenolphthalein, a synthetic red dye, turns colourless in the presence of acid. Phenolphthalein (represented as H-phen in the equation below) is itself slightly acidic:

$$H\text{-phen(aq)} \rightleftharpoons H^+(aq) + phen^-(aq)$$
$$\text{(colourless)} \qquad\qquad \text{(bright red)}$$

The addition of an acid (i.e. more $H^+(aq)$ ions) to this equilibrium will make it, by Le Chatelier's principle, do the opposite and move in the $\leftarrow$ direction. The addition of acid therefore makes the bright-red ions react with $H^+(aq)$ ions to form colourless molecules.

9.5 ACIDS AND THE PERIODIC TABLE

What sort of chemicals react with water to form acids? A few examples are given now.
  (i) Sulphur dioxide

$$SO_2(g) + H_2O(l) \longrightarrow H_2SO_3(aq)$$
$$\text{sulphurous acid}$$

(ii) Carbon dioxide

$$CO_2(g) + H_2O(l) \longrightarrow H_2CO_3(aq)$$
$$\text{carbonic acid}$$

(iii) Nitrogen dioxide

$$2NO_2(g) + H_2O(l) \longrightarrow HNO_3(aq) + HNO_2(aq)$$
$$\text{nitric acid} \qquad \text{nitrous acid}$$

(iv) Phosphorus(V) oxide

$$P_2O_5(s) + 3H_2O(l) \longrightarrow 2H_3PO_4(aq)$$
$$\text{phosphoric(V) acid}$$

These are all oxides of *non-metals*.

Non-metallic oxides show acidic properties

9.6 BASES, ALKALIS AND SALTS

Section 9.1(iv) showed how acidity could be destroyed, or *neutralised*, by adding a metal oxide or a metal hydroxide. Destroying acidity means removing $H^+(aq)$ ions, as illustrated in the following examples.

(a) Copper(II) oxide with hydrochloric acid

When black copper(II) oxide is stirred into a beaker of dilute hydrochloric acid, it dissolves slowly to form a blue-green solution of copper(II) chloride:

$$2HCl(aq) + CuO(s) \longrightarrow CuCl_2(aq) + H_2O(l)$$
$$\text{ionic} \qquad\qquad\qquad \text{ionic}$$

Since Cl^- and Cu^{2+} ions are present at the start and finish of the reaction, they might as well be excluded from the equation. The equation is then written more simply as

$$2H^+(aq) + O^{2-}(s) \longrightarrow H_2O(l)$$

(b) Sodium hydroxide with nitric acid

When a solution of sodium hydroxide is added slowly to dilute nitric acid, the mixture becomes noticeably hotter as the two react exothermically to produce a solution of sodium nitrate:

$$HNO_3(aq) + NaOH(aq) \longrightarrow NaNO_3(aq) + H_2O(l)$$
$$\text{ionic} \qquad\quad \text{ionic} \qquad\quad \text{ionic} \qquad\quad \text{covalent}$$

NO_3^- and Na^+ ions are present at the start and finish of the reaction, and may therefore be omitted from the equation:

$$H^+(aq) + OH^-(aq) \longrightarrow H_2O(l)$$

These two examples show that acidity is neutralised by the reaction of $H^+(aq)$ ions with either O^{2-} or OH^- ions to form water. The acid's negative ions and the metal positive ions left behind comprise a *salt*. In the above examples, copper(II) chloride, $CuCl_2$, and sodium nitrate are *salts*.

In fact, any ionic compound in which the positive metal ion is associated with the negative ion of an acid is called a salt.

Metal oxides and hydroxides are called *bases*, because they are substances which neutralise acids to form salt and water only:

$$acid + base \longrightarrow salt + water$$

A few bases, notably sodium hydroxide and potassium hydroxide, are soluble in water. An aqueous solution of a base is called an *alkali*. Just as $H^+(aq)$ is the cause of acidity, so $OH^-(aq)$ is the cause of alkalinity.

Calcium and lithium hydroxides also dissolve to some extent in water to give alkaline solutions.

In the same way that acids give characteristic colours to certain dyes called indicators (see Sections 9.1(iii) and 9.4(c)), so alkalis tend to give different colours to those same indicators (see Table 9.2).

Table 9.2 *colours of some indicators in acid, alkali and neutral conditions*

Indicator	Colour in acid	Colour when neutral	Colour in alkali
Methyl orange	red	orange	yellow
Phenolphthalein	colourless	colourless	red
Litmus	red	purple	blue
Universal indicator	red	green	purple

9.7 TITRATIONS

Indicators are used to determine the concentrations of acidic or alkaline solutions. The complete operation is known as a *titration*. As an example the titration of dilute hydrochloric acid with sodium hydroxide solution is outlined below, using methyl orange as the indicator.

(i) A pipette is used to measure a precise volume (10 cm³ (10 cubic centimetres) in this case) of the alkali into a conical flask (Fig. 9.4). The base of the meniscus should just touch the graduation mark as shown, and

fig 9.4 *measuring precise amount of liquid with a pipette*

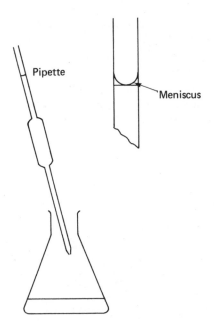

fig 9.5 *arrangement of apparatus for a titration*

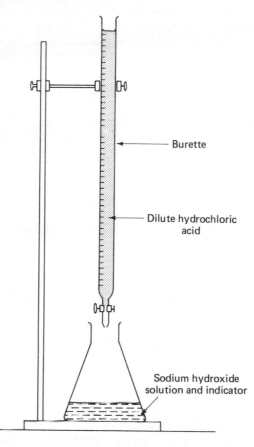

Burette

Dilute hydrochloric acid

Sodium hydroxide solution and indicator

the exact volume is then dispensed by touching the end of the pipette onto the glass side of the flask, *not* by blowing the last few drops out.

(ii) A burette is rinsed out with a little of the acid solution to be used, then filled to the 0 cm³ graduation with the acid solution (Fig. 9.5). Again the base of the meniscus is adjusted so that it just touches the mark.

(iii) A few drops of methyl orange indicator are added to the alkali in the flask, giving the yellow alkaline colour (see Table 9.2). Swirling the flask, acid is run in from the burette in small portions until the neutral orange colour appears. The volume of hydrochloric acid used is then recorded from the burette, and the process repeated until consecutive results agree.

If the concentration of the alkali is known, then the results of such a titration can be used to determine the concentration of the acid. The more dilute the acid, the greater the volume of it required to neutralise the alkali. Details of such calculations are given in Part IV of this book.

Titrations are used to determine the amount of acid or alkali pollution in water, to measure the concentration of ascorbic acid (vitamin C) in orange juice, to determine the chlorine concentration of a bleach, and so on.

Titrations can also be used to prepare salts. In the above example, the salt sodium chloride is obtained:

$$HCl(aq) + NaOH(aq) \longrightarrow NaCl(aq) + H_2O(l)$$

A solid sample of this salt could then be obtained by partially evaporating the solution, then leaving it to crystallise and finally filtering off the crystals (see Section 4.2).

9.8 POLYPROTIC ACIDS AND ACID SALTS

One mole of aqueous hydrogen chloride produces one mole of $H^+(aq)$ ions:

$$HCl(aq) \longrightarrow H^+(aq) + Cl^-(aq)$$

One mole of aqueous sulphuric acid produces two moles of $H^+(aq)$ ions:

$$H_2SO_4(aq) \longrightarrow 2H^+(aq) + SO_4{}^{2-}(aq)$$

One mole of aqueous phosphoric(V) acid produces three moles of $H^+(aq)$ ions:

$$H_3PO_4(aq) \longrightarrow 3H^+(aq) + PO_4{}^{3-}(aq)$$

Hydrochloric acid is therefore a *monoprotic* acid. Sulphuric acid is therefore a *diprotic* acid. Phosphoric(V) acid is therefore a *triprotic* acid.

When one mole of a diprotic acid is neutralised by a base, there are two stages in the neutralisation.

(i)

$$H^+(aq)$$
$$SO_4{}^{2-}(aq)$$
$$H^+(aq)$$

Neutralising both moles of $H^+(aq)$ would require 2 moles of NaOH(aq):

$$H_2SO_4(aq) + 2NaOH(aq) \longrightarrow Na_2SO_4(aq) + 2H_2O(l)$$
$$\text{sodium}$$
$$\text{sulphate}$$

Crystals of sodium sulphate may be obtained by completely neutralising the acid with sodium hydroxide solution in a titration similar to that described in Section 9.7.

(ii)

$$H^+(aq)$$
$$SO_4{}^{2-}(aq)$$
$$H^+(aq)$$

Neutralising only one mole of $H^+(aq)$ would require 1 mole of NaOH(aq):

$$H_2SO_4(aq) + NaOH(aq) \longrightarrow NaHSO_4(aq) + H_2O(l)$$

sodium
hydrogen
sulphate

Sodium hydrogen sulphate is an *acid salt*, since a solution of it in water contains $H^+(aq)$ ions. Titrating sulphuric acid with sodium hydroxide solution produces sodium sulphate. Crystals of sodium hydrogen sulphate may be obtained by adding the same volume of sulphuric acid needed in the titration to exactly *half* the volume of sodium hydroxide. Only half the $H^+(aq)$ ions in the acid would now be neutralised, and evaporation followed by crystallisation of the concentrated solution would now produce crystals of the acid salt.

The triprotic acid phosphoric(V) acid, H_3PO_4, has three different sodium salts:

Na_3PO_4	sodium phosphate(V)	(normal salt)
Na_2HPO_4	disodium hydrogen phosphate(V)	(acid salt)
NaH_2PO_4	sodium dihydrogen phosphate(V)	(acid salt)

9.9 HOW ACID? THE pH SCALE

Different degrees of acidity may be obtained by diluting acidic solutions, for example as shown in the tubes in Fig. 9.6. The decrease in $H^+(aq)$ ion

fig 9.6 *the reaction of magnesium ribbon with different concentrations of hydrochloric acid*

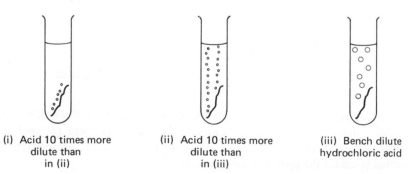

(i) Acid 10 times more
dilute than
in (ii)

(ii) Acid 10 times more
dilute than
in (iii)

(iii) Bench dilute
hydrochloric acid

concentration in going from tube (iii) to tube (i) is shown by the decrease in vigour of the reaction with magnesium. The solution in tube (i) is less acidic than that in tube (iii). However, merely observing the rate of bubbling after adding magnesium ribbon makes for a rather inaccurate means of measuring acidity.

Acidity is frequently measured using *universal indicator*. By mixing together various different indicators which change colour at slightly different acidities, chemical manufacturers have developed a universal indicator which goes through a spectrum of colours as the acidity changes. These colours can be seen in the following experiments (see Fig. 9.7).

Acidity is measured on a scale from 0 (very acid) to 14 (very alkaline). This is the pH scale, where pH stands for 'hydrogen *potenz*' (hydrogen power) and measures the $H^+(aq)$ concentration. In 1 dm^3 (one cubic decimetre) of pure water there are 0.000 0001 or 10^{-7} moles of $H^+(aq)$ ions. The pH of pure water is 7. In a highly acidic solution there may be 0.1 or 10^{-1} moles of $H^+(aq)$ per dm^3. For this solution the pH would be 1.

The pH scale should be thought of simply as numbers which indicate acidity. pH numbers are related to universal indicator colours as follows:

red	orange	yellow	green	blue	deep blue	purple
0	3	6	7	8	10	14

The pH of a solution can be determined by adding a few drops of universal indicator solution, then reading off the pH number from the corresponding colour. Colour charts are provided by the manufacturer for this purpose. For example, tube (i) in the experiment in Fig. 9.6 gave a yellow colour on adding a few drops of universal indicator. Its pH is therefore 6, corresponding to only slight acidity. Paper impregnated with universal indicator is often more conveniently used in place of the solution itself.

For more precise measurements, pH meters are available. These are used with a delicate glass electrode which, when dipped into a solution, will give a direct reading of pH on the meter.

9.10 WEAK AND STRONG ACIDS

Solution a	*Solution b*	*Solution c*
Phenol in	Ethanoic acid	Hydrochloric acid
water	in water	in water

Solutions a, b and c contain identical concentrations of three different acids: phenol (used in many disinfectants), ethanoic acid (present in vinegar) and hydrochloric acid. Addition of a few drops of universal indicator to each tube produces the following colours:

fig 9.7 *the colours of universal indicator*

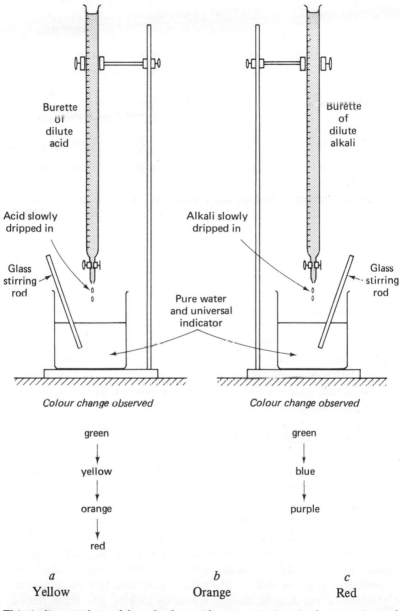

This indicates that, although the acid concentration is the same in each tube, the acidities are not. Tube *a* obviously has a much lower concentration of $H^+(aq)$ ions than tube *c*. This is confirmed by a comparison of the ability of the three solutions to conduct electricity:

a	b	c
Just allows current to flow	Allows a moderate current to flow	Conducts very well

Molecules of phenol (tube a) are providing fewer H^+ ions than molecules of hydrogen chloride (tube c). This suggests varying degrees of ionisation.

phenol $\qquad$ $C_6H_5OH(aq) \rightleftharpoons C_6H_5O^-(aq) + H^+(aq)$

ethanoic acid $\qquad$ $CH_3CO_2H(aq) \rightleftharpoons CH_3CO_2^-(aq) + H^+(aq)$

hydrochloric acid $\quad ClH(aq) \longrightarrow Cl^-(aq) \qquad + H^+(aq)$

Hydrochloric acid is described as a *strong acid*: it is ionised completely in solution.

Ethanoic acid and phenol are *weak acids*: they are only partially ionised in solution. Furthermore, experiment suggests that phenol is a weaker acid than ethanoic acid.

9.11 DANGEROUS ACIDS

It is true that all strong acids can be dangerous, because a high concentration of $H^+(aq)$ can, for example, break down large protein molecules of our skin, destroy cells and react with important body chemicals to make them useless. However, many acids are dangerous for other reasons. Nitric acid, in addition to being an acid, is also a strong oxidising agent, oxidising organic matter including ourselves to steaming lumps of carbon. Concentrated sulphuric acid is a strong dehydrating agent, removing water from many organic substances to leave a charred mass.

9.12 ACIDS IN EVERYDAY LIFE

Sulphuric acid is by far the most important acid of the consumer society, being essential to the manufacture of most dyes, plastics, explosives, steel, fertilisers, paints and drugs.

The digestive juices in our stomachs break up food only in acidic conditions, and our stomachs are therefore maintained at a pH of between 1 and 2 by the presence of hydrochloric acid. Too much acidity in the stomach is neutralised by swallowing 'health salts' containing mainly sodium hydrogen carbonate together with a little solid tartaric acid. In water these react to produce carbon dioxide, causing a healthy looking fizz. When the carbonate meets the hydrochloric acid in the stomach more fizzing occurs, excess acid is neutralised, the victim 'burps' and his indigestion is relieved.

The acid tastes of grapefruit, lemon, oranges and vinegar are due to tartaric, citric or ethanoic acids. The protein we eat (meat, fish, nuts) or wear (silk, wool) is made up of a variety of amino acids, and a major constituent of fat is stearic acid. The taste of butter, cheese and some margarines is due in part to traces of strong smelling butanoic and pentanoic acids.

Even the skin holding us together is made up of long chains of amino acids. It is certainly true that acids are a big part of our lives.

SUMMARY OF CHAPTER 9

Acids are substances which
  taste sour,
  react with metals high in the electrochemical series to give hydrogen,
  generally react to give carbon dioxide with carbonates,
  change the colour of certain dyes called *indicators*.
Acids are *neutralised* by bases, *salts* being formed in the process.

Acidity results from the presence of the $H^+(aq)$ ion. This ion is liberated by acids only in the presence of water.

Strong acids dissociate completely in aqueous solution to produce $H^+(aq)$ ions. Those which dissociate only partially are weak acids.

Soluble bases are alkalis. Alkalinity results from the presence of the $OH^-(aq)$ ion.

Monoprotic acids can liberate only one mole of $H^+(aq)$ ions per mole, whilst one mole of di- or triprotic acids can liberate two or three moles of $H^+(aq)$ ions, respectively. The partial neutralisation of such polyprotic acids produces *acid salts*.

The degree of acidity or alkalinity of an aqueous solution is measured on the pH scale. The more acidic the solution the lower its pH value; the more alkaline the higher its pH, with a pH of 7 indicating a neutral solution.

128

PART III
CHEMICAL BEHAVIOUR

There are millions upon millions of chemicals each containing different combinations and arrangements of the one hundred and five known elements. A *bad way* of studying chemistry would be separately to learn the properties of each one. A more rational approach is to sort chemicals into classes of similar behaviour. We have already seen how a large number of chemicals with specific properties in common have been classified as acids.

The most widely used classification of chemicals is the Periodic Table of the elements (see Section 2.6). This section looks at the properties of the more common elements and their compounds in relation to their place in the Periodic Table. It deals with experimental observations—with facts. Frequent reference will be made to the theory of electrons rushing around positive nuclei (see Part I) because this picture seems to fit the facts quite well. However, it is important in any science to realise that some day someone will suggest a better picture: a new theory will emerge The experimental facts of this section will then still hold true: TNT will still explode when struck, sodium will still react with water, lead will still be a dense metal. It is only our explanations that are open to change.

PART III

CHEMICAL BEHAVIOUR

PREDICTIONS AND TRENDS

10.1 THE PERIODIC TABLE

Vertical columns of elements in the Periodic Table are called *groups* (see Section 2.6). In addition there are the transition metals found between groups 2 and 3. Elements within the same group show many similar properties. An efficient way of studying the elements and their compounds is therefore to consider them by group. The positions in the Table of the more common elements considered in the following chapters are shown in Table 10.1.

An examination of some changes going across, up or down the Table will help in relating chemical behaviour to the theories outlined in Part I.

Table 10.1   *the Periodic Table, showing the position of elements to be studied*

1	2				3	4	5	6	7	0
Li	Be				B	C	N	O	F	Ne
Na	Mg	transition metals			Al	Si	P	S	Cl	Ar
K	Ca		Fe	Cu	Zn	Ge			Br	Kr
									I	Xe

(a) Atomic size

Going from left to right across the Periodic Table, the number of outer electrons increases, but nevertheless they are all in the same energy level, all a similar distance from the nucleus (Fig. 10.1).

Going from left to right, the positive charge on the nucleus increases. Thus the outer electrons of argon are pulled towards the nucleus by an

fig 10.1 *atomic size across the Periodic Table*

18+ charge, whereas the outer electron of sodium is pulled only by an 11+ charge. Atoms of argon are therefore smaller than those of sodium. Hence

Atomic size decreases

Nuclear charge also increases moving down the Table, but now outer electrons are in successively more distant energy levels (Fig. 10.2). The

fig 10.2 *atomic size down the Periodic Table*

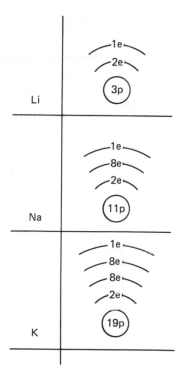

nuclear charge is *screened* (i.e. protected) from the outer electrons by a greater number of filled inner energy levels. These inner electrons are called *screening electrons*. Hence

Atomic size increases

(b) Electronegativity and electropositivity

A smaller atom with fewer screening electrons will attract nearby electrons more strongly. If the number of screening electrons is the same, then nearby electrons will be attracted more strongly towards whichever atom has the greater nuclear charge. The tendency of an atom to pull electrons towards it is called its *electronegativity*. Fluorine is more *electronegative* than either chlorine or oxygen because fluorine shows a stronger pull for electrons (Fig. 10.3). Hence

Electronegativity increases

Electronegativity increases

fig 10.3 *electronegativity*

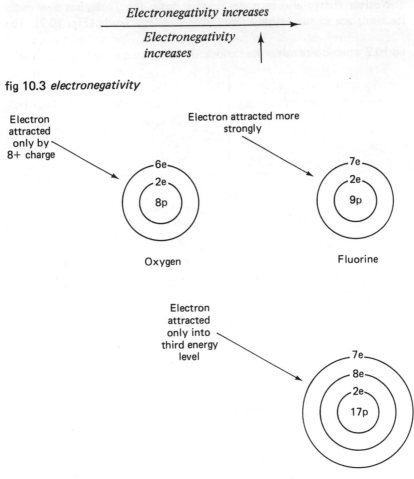

Similarly electropositivity, the tendency to lose electrons, changes as shown below:

Electropositivity increases

Electropositivity increases

From their positions in the Periodic Table, it can be seen that potassium will be more electropositive than either sodium or calcium.

(c) Bonding

The outer electrons of electropositive atoms can be easily lost into a de-localised electron sea. Such atoms therefore bond together in giant metallic structures (see Sections 3.6 and 3.7). These are the metals, the electropositive elements found towards the left and bottom of the Periodic Table.

The non-metals join to one another through covalent bonds, usually forming molecular structures (see Section 3.8).

When metals (electropositive) bond with non-metals (electronegative), ionic compounds are formed (see Section 3.3(f)).

These three situations are shown in Fig. 10.4.

fig 10.4 *bonding between atoms*

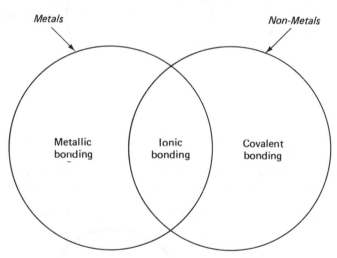

10.2 POLARISATION

Metals react with non-metals to form ionically bonded compounds. In 1924 Fajans and others proposed that the size and charge of ions might influence the properties of the resulting compound. Fig. 10.5 shows the relative sizes and charges on a number of common positive ions. Negative ions are in general larger.

A very small positive ion can get especially close to a negative ion, attracting and dragging outer electrons from it. If the small cation has a high positive charge, this distortion of electrons will be even more pronounced. Thus an Fe^{3+} ion will distort or *polarise* a negative ion, pulling electrons from it (Fig. 10.6).

fig 10.5 *the relative sizes of some ions*

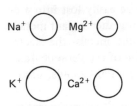

Li⁺ $\bigcirc$

Na⁺ $\bigcirc$ Mg²⁺ $\bigcirc$ Al³⁺ $\bigcirc$

K⁺ $\bigcirc$ Ca²⁺ $\bigcirc$ $\bigcirc$ Fe³⁺ $\bigcirc$ Cu²⁺

fig 10.6 *negative ion distorted by polarising positive ion*

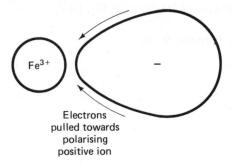

Fe³⁺

Electrons
pulled towards
polarising
positive ion

In contrast, a potassium ion would show very little of this *polarising* effect (Fig. 10.7).

Small and/or highly charged positive ions are therefore *polarising* ions. For example Al^{3+}, Fe^{3+} and Cu^{2+} are polarising ions. K^+ is non-polarising.

fig 10.7 *negative ion unaffected by non-polarising positive ion*

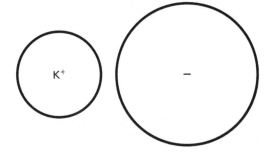

K⁺

There are three consequences of polarisation.

(a) Degree of covalency

Through polarisation, electrons are pulled to a position between the positive and negative ions (see Fig. 10.6). It is as though the two ions are partially sharing electrons. Thus the ionic compound has a *degree of covalency*, and may show some covalent properties. For example, aluminium chloride, in which the chloride ion is polarised by the small Al^{3+} ion, vaporises when heated gently, just like a truly covalent volatile solid. Iron(III) chloride behaves similarly.

(b) Decomposition of negative ions

Ions are stable because they have achieved the inert-gas electron configuration. A negative ion will therefore become less stable if electrons are distorted away from it. The polarisation of a large negative ion may cause it to decompose to one which is smaller and less easily distorted. For example

$$
\left.
\begin{array}{l}
OH^- \text{ (hydroxide)} \\
CO_3^{2-} \text{ (carbonate)} \\
NO_3^- \text{ (nitrate)}
\end{array}
\right\}
\text{all decompose to the smaller } O^{2-} \text{ (oxide) ion}
$$

Thus most carbonates and hydroxides decompose to the oxide on heating. For example

$$Mg(OH)_2(s) \longrightarrow MgO(s) + H_2O(l)$$
$$CaCO_3(s) \longrightarrow CaO(s) + CO_2(g)$$

However, potassium and sodium ions, being non-polarising, leave their carbonates and hydroxides unaffected by heat.

(c) Hydration

In addition to dragging electrons from negative ions, polarising positive ions will also attract any negative charge. Thus polarising ions attract the negative ends of water molecules around them, and salts containing these ions will be strongly *hydrated*. For example

iron(III) nitrate	$Fe(NO_3)_3.9H_2O$
aluminium sulphate	$Al_2(SO_4)_3.18H_2O$
magnesium chloride	$MgCl_2.6H_2O$

Salts of potassium, a non-polarising ion, are rarely hydrated.

SUMMARY OF CHAPTER 10

In the Periodic Table

(i) atomic size and electropositivity increase ← ↑ ,
(ii) electronegativity increases → ↑ ,
(iii) metals (to the left in the Table) bond with each other through metallic bonding,

(iv) non-metals (to the right in the Table) bond with each other through covalent bonding, and

(v) metals tend to bond with non-metals through ionic bonding.

A small, highly charged positive ion is said to be polarising. Such an ion tends

(i) to hydrate more easily,

(ii) to decompose large negative ions with which it is combined, and

(iii) to form chlorides which are more volatile.

CHAPTER 11

THE METALS

11.1 GENERAL PROPERTIES OF METALS

The *metals* that we shall consider are shown in *italics* in Table 11.1.

In order to form metallic bonds with each other, the atoms must *lose* outer electrons into a spread-out delocalised electron cloud. Metallic bonding is continuous from one atom to the next, creating a giant structure. The smaller the atoms the stronger the metallic bonds between them. The atoms of most metals are relatively small, so in general the metals are non-volatile (not easily vaporised) with high melting points and high heats of fusion (see Section 5.2). The large size of the group 1 atoms accounts for the softness and low melting points of these metals.

Table 11.1 *the metals*

1	2				3	4	5	6	7	0
Li	*Be*				B	C	N	O	F	Ne
Na	*Mg*	*transition elements*			*Al*	Si	P	S	Cl	Ar
K	*Ca*		*Fe*	*Cu* *Zn*	*Ge*				Br	Kr
									I	Xe

Metallic bonds draw atoms together in any direction. If hit by a hammer, the metal atoms are pushed into a new shape. Since they are all equally well held to one another in this new arrangement, the atoms stay put. Thus a solid metal can be dented—it is *malleable* (Fig. 11.1).

For a similar reason, metals can be pulled out into long wires—a metal is *ductile* (Fig. 11.2).

Metals have freely moving delocalised electrons. For this reason, metals are good conductors.

fig 11.1 *a metal is malleable*

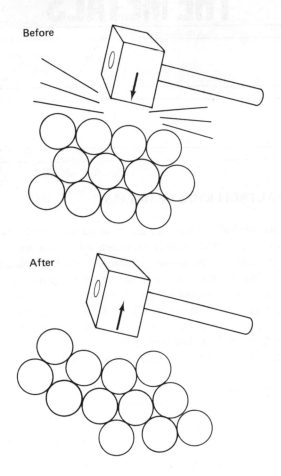

Before

After

11.2 GROUP 1, THE ALKALI METALS

The *alkali metals* are shown in *italics* in Table 11.2, and are listed below.

lithium, Li   3p 2e 1e
sodium, Na   11p 2e 8e 1e
potassium, K   19p 2e 8e 8e 1e
rubidium, Rb
caesium, Cs

(a) General properties
Being at the extreme left of the Periodic Table, atoms of the alkali metals

fig 11.2 *a metal is ductile*

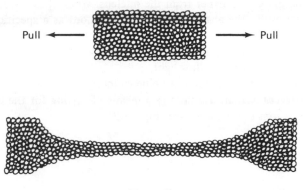

Pull ◄─────   ─────► Pull

After pull

Table 11.2 *the alkali metals*

1	2					3	4	5	6	7	0
Li	Be					B	C	N	O	F	Ne
Na	Mg	transition metals				Al	Si	P	S	Cl	Ar
K	Ca		Fe	Cu	Zn		Ge			Br	Kr
Rb										I	Xc
Cs											

are large and highly electropositive. Each atom will lose one electron easily

$$M \longrightarrow M^+ + 1e$$

to leave a singly charged cation with the stable inert-gas structure.

Being large atoms, the metallic bonding is weak, getting weaker from lithium to potassium. These metals

(i) are white, shiny solids,

(ii) conduct electricity and heat well, and are malleable and ductile

because they are metallically bonded

(iii) are soft (they can be cut with a knife), with low melting points and low heats of fusion

because the metallic bonding is weak

Potassium, for example, melts at 63°C, compared with iron at 1539°C.

When ions of these metals are given energy, electrons are excited up to higher energy levels further from the nucleus. When these electrons jump back down again, the absorbed energy is given out as a specific colour of light:

lithium ions give out crimson light,

sodium ions give out yellow light,

potassium ions give out a misty blue-violet light.

These different colours are used as a means of testing for the presence of alkali metal ions in the *flame test* (Fig. 11.3).

fig 11.3 *the flame test*

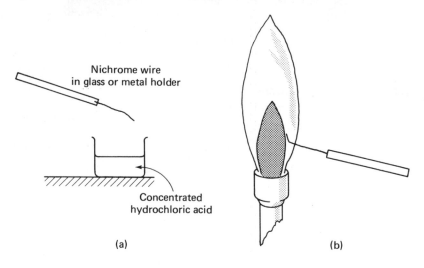

(a)

(b)

Nichrome wire in glass or metal holder

Concentrated hydrochloric acid

The flame test

  (i) The nichrome wire is cleaned by dipping it into concentrated hydro-chloric acid (Fig. 11.3a).

 (ii) The tip of the wire is held in the side of a colourless gentle bunsen flame (Fig. 11.3b). If clean, no strong colour is seen in the flame.

(iii) Again the wire is dipped into concentrated hydrochloric acid, then touched onto a small sample of the test solid.

 (iv) Again the tip of the wire is held in the flame.

  (v) The flame colour indicates whether Li^+, Na^+ or K^+ is present.

(b) Chemical reactions

In every reaction of the group 1 metals, outer electrons are vigorously given to some other substance. The metals are therefore oxidised (they lose electrons), whilst the substance gaining the electrons is reduced (see Chapter 7). The ease of losing electrons, and therefore the reactivity, increases down the group, so that potassium is more reactive than sodium, which is more reactive than lithium.

All the reactions are strongly exothermic.

(i) *Giving electrons to water*

The alkali metals react with water as shown in Fig. 11.4.

fig 11.4 *the alkali metals with water*

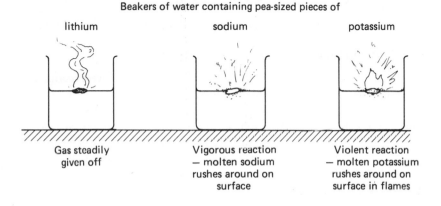

Beakers of water containing pea-sized pieces of

| lithium | sodium | potassium |

| Gas steadily given off | Vigorous reaction – molten sodium rushes around on surface | Violent reaction – molten potassium rushes around on surface in flames |

With lithium it is safe to collect the gas (Fig. 11.5). Quickly removing the tube of gas and holding it to a lighted splint gives a loud 'pop'—a positive test for hydrogen.

Addition of universal indicator to each of the three beakers shown in

fig 11.5 *collecting the gas from lithium's reaction with water*

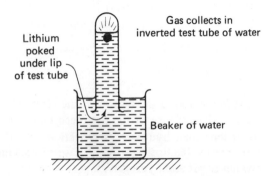

Lithium poked under lip of test tube

Gas collects in inverted test tube of water

Beaker of water

Fig. 11.4 shows a strongly alkaline solution, suggesting OH^- ions to be present (see Section 9.6):

$$2Na(s) + 2H_2O(l) \longrightarrow 2NaOH(aq) + H_2(g)$$

(ii) *Giving electrons to chlorine gas*

When a combustion spoon holding a small piece of sodium is lowered into a gas jar of chlorine, the sodium burns brightly giving an off-white smoke of tiny sodium chloride particles. Chlorine has gained electrons to form chloride ions, Cl^-:

$$2Na(s) + Cl_2(g) \longrightarrow 2NaCl(s)$$

As expected, the reaction is faster for potassium and slower for lithium.

(iii) *Giving electrons to oxygen gas*

Using a gas jar of oxygen in place of the chlorine in the above experiment, a similar smoke and bright-yellow spluttering flame is seen as the oxides are formed. Lithium burns with a crimson light and potassium with a misty blue light. If, after reaction, a few drops of universal indicator solution are shaken in the gas jar, an intense purple colour is obtained, supporting the statement of Section 9.6 that metallic oxides are basic.

Lithium forms the simple oxide, in which oxygen has gained electrons to form O^{2-} ions:

$$4Li(s) + O_2(g) \longrightarrow 2Li_2O(s)$$

Sodium and potassium form more complex oxides as well as the normal oxides Na_2O and K_2O:

$$2Na(s) + O_2(g) \longrightarrow Na_2O_2(s)$$
$$\text{sodium } peroxide$$

Sodium peroxide contains the O_2^{2-} peroxide ion.

$$K(s) + O_2(g) \longrightarrow KO_2(s)$$
$$\text{potassium } superoxide$$

Potassium superoxide contains the O_2^- superoxide ion.

(iv) *Giving electrons to acids*

Since group 1 metals are found at the top of the electrochemical series, they ionise readily at the expense of the hydrogen ions in an acid, being more favourable as ions than hydrogen (see Section 9.4(a)). The metals therefore react vigorously (explosively in the case of sodium and potassium) to form the metal salt and hydrogen gas:

$$2Na(s) + 2HNO_3(aq) \longrightarrow 2NaNO_3(aq) + H_2(g)$$
sodium
nitrate

(v) *Giving electrons to the air*

The group 1 metals tarnish rapidly in air, giving electrons to oxygen, water vapour and then carbon dioxide to form a powdery white coating of the metal carbonate. To prevent their reaction with air, the metals are kept under paraffin oil.

(c) Compounds of the group 1 metals

Li^+, Na^+ and K^+ ions are colourless. Compounds of these metals are therefore white or transparent. They are ionic crystalline solids which, with the exception of lithium hydroxide and lithium carbonate, dissolve well in water.

Again with the exception of lithium, which has a relatively small ion, the large singly charged alkali metal ions are non-polarising (see Chapter 10).

(i) *Oxides*

The oxides react with water exothermically to form strongly alkaline solutions of the hydroxide:

$$Li_2O(s) + H_2O(l) \longrightarrow 2LiOH(aq)$$

The oxides of sodium and potassium hold excess oxygen which is liberated in the reaction:

$$2Na_2O_2(s) + 2H_2O(l) \longrightarrow 4NaOH(aq) + O_2(g)$$

$$4KO_2(s) + 2H_2O(l) \longrightarrow 4KOH(aq) + 3O_2(g)$$

(ii) *Hydroxides*

The hydroxides dissolve in water to form strongly alkaline solutions. Concentrated solutions of the hydroxides are very corrosive, causing burns and blistering to the skin. Pellets of the solids attract water from the air so strongly that they form small pools of corrosive solution within an hour.

A substance which attracts water in which it then dissolves is said to be *deliquescent*.

Only lithium hydroxide decomposes when heated, since only the lithium ion is sufficiently polarising (see Section 10.2):

$$2LiOH(s) \longrightarrow Li_2O(s) + H_2O(l)$$

Sodium and potassium hydroxides are stable to heat.

(iii) *Other compounds*

Lithium carbonate and lithium nitrate similarly decompose to the oxide on heating:

$$Li_2CO_3(s) \longrightarrow Li_2O(s) + CO_2(g)$$

$$4LiNO_3(s) \longrightarrow 2Li_2O(s) + 4NO_2(g) + O_2(g)$$

The decomposition of lithium nitrate is shown in Fig. 11.6.

fig 11.6 *the decomposition of lithium nitrate*

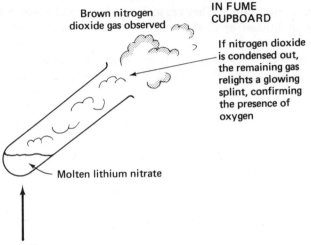

Brown nitrogen dioxide gas observed

IN FUME CUPBOARD

If nitrogen dioxide is condensed out, the remaining gas relights a glowing splint, confirming the presence of oxygen

Molten lithium nitrate

HEAT

The large singly charged non-polarising sodium and potassium ions render sodium carbonate and potassium carbonate stable to heat. However, sodium nitrate and potassium nitrate are partially decomposed by heat, losing only oxygen to form the nitrite ion, NO_2^-:

$$2KNO_3(s) \longrightarrow 2KNO_2(s) + O_2(g)$$

Thus when potassium nitrate is heated, no brown fumes of nitrogen dioxide are obtained.

(iv) *Hydration*

Only lithium ions are sufficiently polarising to show an appreciable attraction for water molecules. Thus lithium salts show a high degree of *hydration*.

(d) The group 1 metals in everyday life

Lithium compounds provide the brilliant crimson light of many fireworks. Lithium chloride and lithium fluoride are added to glass to control the melting point and to reduce the chance of cracking when subjected to extremes of temperature.

The nuclear pile of an atomic power station generates sufficient energy to destroy the entire reactor in seconds. It is essential that heat is conducted away rapidly, and the low melting point and high conductivity of sodium and potassium make them ideal for this purpose. A molten mixture of the two metals is circulated rapidly through the pile, transferring heat to conventional steam turbines which generate the electricity.

A trace of sodium is responsible for the yellow glow of sodium-vapour street lights (Fig. 11.7). Such lights are very efficient in their use of electricity, with the added advantage that yellow is a colour which is not easily scattered and lost through a mist or fog.

Sodium chloride and a little potassium chloride are essential to life. The very process by which instructions are transmitted to and from the brain involves a rapid exchange of Na^+ and K^+ ions along the nerves.

Sodium hydroxide is sold in tins in hardware stores as 'caustic soda'. It is used in the home to clean out drains because of its ability to attack organic dirt and grease to produce water-soluble soaps (see Section 14.2(a)). Industrially, caustic soda is boiled with animal fat or vegetable oil to make soap, or with wood pulp to dissolve our resinous products such as lignin, leaving cellulose fibres from which paper is made.

A mixture of caustic soda and carbon disulphide dissolves cellulose to form an orange gelatinous liquid. When injected through a set of spinning holes (a *spinnerette*) into a bath of dilute acid, the cellulose precipitates as a fine-spun thread of artificial silk, or *rayon*.

fig 11.7 *a sodium-vapour light*

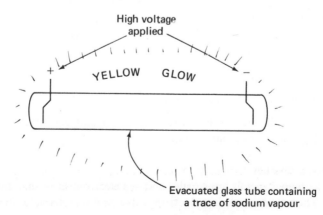

High voltage applied

YELLOW GLOW

Evacuated glass tube containing a trace of sodium vapour

Sodium carbonate acts as a mild alkali. It, too, attacks dirt and grease to form water-soluble products, but it is much safer to use and less corrosive. It is sold as 'washing soda crystals', $Na_2CO_3.10H_2O$, in which more than half the weight carried home by the shopper is water. Once exposed to the air, the transparent crystals lose water of hydration (they *effloresce*) to leave a white powder of $Na_2CO_3.1H_2O$.

Plants cannot grow unless potassium ions are present in the soil. Many fertilisers therefore contain potassium salts. Wood ashes are often scattered around rose bushes because of their high potassium content. On a larger scale, vast deposits of potassium nitrate are excavated each year and used directly as fertiliser. Potassium nitrate is also used in gunpowder to provide oxygen for the explosive combustion of carbon and sulphur.

11.3 GROUP 2, THE ALKALINE EARTH METALS

The *alkaline earth metals* are shown in *italics* in Table 11.3, and are listed below.

Table 11.3   *the alkaline earth metals*

1	2				3	4	5	6	7	0
Li	*Be*				B	C	N	O	F	Ne
Na	*Mg*	transition metals			Al	Si	P	S	Cl	Ar
K	*Ca*	Fe	Cu	Zn	Ge				Br	Kr
	Sr								I	Xe
	Ba									
	Ra									

```
beryllium, Be     4p 2e 2e
magnesium, Mg    12p 2e 8e 2e
calcium, Ca      20p 2e 8e 8e 2e
strontium, Sr
barium, Ba
radium, Ra
```

This chapter deals principally with magnesium and calcium, the most commonly encountered group 2 metals.

(a) General properties

Atoms of group 2 metals are smaller and less electropositive than those of group 1. They are nevertheless electropositive, and this character increases down the group. The atoms achieve an inert-gas electron configuration by the loss of two electrons:

$$M \longrightarrow M^{2+} + 2e$$

Since the atoms are smaller, group 2 metals are bonded more strongly than those of group 1. These metals
 (i) are shiny solids,
 (ii) conduct electricity and heat well, and are malleable and ductile

because they are metallically bonded

(iii) are harder, with higher melting points and heats of fusion than the group 1 metals

because the metallic bonding is stronger than for group 1

Calcium, for example, melts at 850°C, compared with potassium at 63°C.
 Magnesium ions give no distinctive colour in the flame test, but calcium ions produce a faint orange-red coloration specific to that metal.

(b) Chemical reactions
Every reaction of these metals involves the two outer electrons being given to some other substance. The metals are therefore oxidised (loss of electrons) during reaction, whilst the substance gaining the electrons is reduced (see Chapter 7). Calcium, being the more electropositive, is more reactive than magnesium.

(i) *Giving electrons to water*

When dropped into a beaker of water, the metals sink. Magnesium reacts very slowly to form small bubbles of hydrogen which stick to its surface. A steady stream of bubbles rises from calcium and can be collected in an inverted test tube filled with water. When a flame is held to the tube, the 'pop' shows the gas to have been hydrogen. The resulting cloudy suspension in the beakers is faintly alkaline, turning universal indicator solution blue:

$$Ca(s) + 2H_2O(l) \longrightarrow Ca(OH)_2(aq) + H_2(g)$$
$$\text{slightly soluble}$$
$$\text{calcium hydroxide}$$

With steam, even the reaction with magnesium is violent.

(ii) *Giving electrons to chlorine*

When a coil of burning magnesium is lowered into a gas jar of chlorine, the metal burns with a brilliant white light. A fine white ash of magnesium chloride remains:

$$Mg(s) + Cl_2(g) \longrightarrow MgCl_2(s)$$

The reaction with calcium is similar although faster, and with a brilliant orange flame.

(iii) *Giving electrons to oxygen*

The reactions are identical to those with chlorine:

$$2Mg(s) + O_2(g) \longrightarrow 2MgO(s)$$

(iv) *Giving electron to acids*

Being high in the electrochemical series, well above hydrogen, the metals react vigorously with hydrogen ions in dilute acids to form the metal salt and hydrogen (see Section 9.4(a)):

$$Ca(s) + 2HCl(aq) \longrightarrow CaCl_2(aq) + H_2(g)$$
$$\text{calcium}$$
$$\text{chloride}$$

(v) *Tarnishing*

In air, calcium reacts at a steady rate to form the oxide, then the hydroxide and finally white crumbly carbonate. In order to protect it from tarnishing, it is usually kept under paraffin oil.

Although shining white when clean, magnesium surfaces are generally black because of a thin coating of magnesium nitride, Mg_3N_2, which forms by the slow reaction of the metal with nitrogen in the air.

(c) Compounds of the group 2 metals

Group 2 ions are colourless. Compounds of these metals are therefore white or transparent. All are ionic, crystalline solids.

(i) *Oxides*

The oxides react with water to form hydroxides. Reaction rate increases going down the group:

$$MgO(s) + H_2O(l) \longrightarrow Mg(OH)_2(s)$$
$$CaO(s) + H_2O(l) \longrightarrow Ca(OH)_2(s)$$

In the case of calcium oxide, the reaction is highly exothermic.

(ii) *Hydroxides*

The solubility of the hydroxides increases going down the group, magnesium hydroxide being only very sparingly soluble. Solutions of the hydroxides are alkaline because of the presence of OH^-(aq) ions.

A solution of calcium hydroxide in water is 'lime water', which is used to test for carbon dioxide gas. The gas, being acidic, reacts with alkaline calcium hydroxide solution to produce a milky white suspension of calcium carbonate:

$$CO_2(g) + Ca(OH)_2(aq) \longrightarrow CaCO_3(s) + H_2O(l)$$

If more carbon dioxide is bubbled through this suspension, the cloudiness fades as the soluble acid salt, calcium hydrogen carbonate, is formed:

$$CaCO_3(s) + H_2O(l) + CO_2(g) \longrightarrow Ca(HCO_3)_2(aq)$$
calcium   calcium hydrogen
carbonate   carbonate (soluble)
(insoluble)

The presence of this acid salt in hard water accounts for the 'furring up' of pipes, boilers and kettles, the formation of stalactites and stalagmites in caves and the formation of scum when using soap (see Section 12.9(b)).

(iii) *Solubility of salts*

Chloride and nitrate salts of these metals are soluble. Magnesium sulphate is soluble, but the sulphates of calcium and below are only very sparingly soluble. The carbonates are all highly insoluble.

(iv) *Decomposition of compounds*

With their smaller size and double charge, the ions are more polarising than those of group 1. Negative ions are therefore more easily decomposed. Thus the hydroxides, carbonates and nitrates all decompose to the simple oxide on heating:

$$Ca(OH)_2(s) \longrightarrow CaO(s) + H_2O(l)$$

$$MgCO_3(s) \longrightarrow MgO(s) + CO_2(g)$$

$$2Ca(NO_3)_2(s) \longrightarrow 2CaO(s) + 4NO_2(g) + O_2(g)$$

The decomposition of calcium hydroxide is shown in Fig. 11.8, and that of magnesium carbonate in Fig. 11.9.

(v) *Hydration of salts*

The greater polarising power of the group 2 ions compared with those of group 1 is reflected in the degree of hydration of their salts (see Table 11.4)

fig 11.8   *the decomposition of calcium hydroxide*

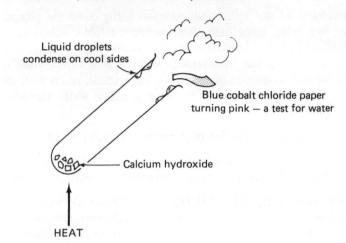

fig 11.9   *the decomposition of magnesium carbonate*

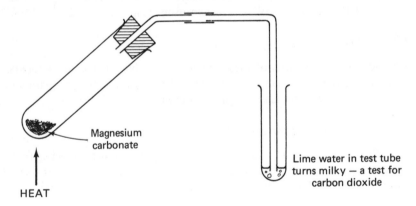

Table 11.4   *the hydration of some group 2 salts*

$MgSO_4.7H_2O$	$CaSO_4.2H_2O$
$MgCl_2.6H_2O$	$CaCl_2.6H_2O$
$Mg(NO_3)_2.6H_2O$	$Ca(NO_3)_2.4H_2O$

Smaller, more polarising, Mg^{2+} ions are generally more strongly hydrated.

If the lids are not replaced tightly on bottles of many magnesium compounds, their marked attraction for the water vapour of the air causes them to change from pure crystals into a concentrated aqueous solution within a few days. This process is *deliquescence.*

Anhydrous calcium chloride absorbs water to form the solid hydrated crystals, $CaCl_2.6H_2O$. A substance which remains solid after taking in water in this way is said to be *hygroscopic.* Solid lumps of calcium chloride are used widely as a drying agent.

(d) The group 2 metals in everyday life

Magnesium has an exceptionally low density. It combines with aluminium to form a strong, unreactive low-density alloy, Almag, used extensively in the aerospace industry.

Since metal oxides are basic (see Section 9.6) and magnesium oxide is non-toxic, a suspension of magnesium oxide in water can be taken as a medicine to neutralise excess acid in the stomach. This is 'milk of magnesia'.

Plaster of Paris is partially hydrated calcium sulphate, $CaSO_4.\frac{1}{2}H_2O$. The addition of water to this powder allows a giant ionic solid lattice of $CaSO_4.2H_2O$ to form, and thus the plaster sets. This absorption of water is accompanied by a slight volume increase, and this serves to push the plaster into the fine detail of a mould.

In the production of iron from iron ore, sandy impurities are removed by the addition of limestone, calcium carbonate. In the heat of the blast furnace, it reacts with the sand to form a molten 'slag' which can be removed easily (see Section 11.8). Vast quantities of limestone are also heated strongly with clay and sand to produce cement (see Section 11.8). Quicklime, calcium oxide, is obtained by the action of heat on limestone. Addition of water then forms the hydroxide, slaked lime, which is mixed with sand to form mortar. Mortar was used by the Romans. Slowly, by reaction with carbon dioxide over hundreds of years, it changes into calcium carbonate. Roman buildings have been dated from the carbonate content of their mortar.

11.4 ALUMINIUM

Aluminium is the only common metal in group 3 of the Periodic Table (Table 11.5), and its electron arrangement is

aluminium, Al   13p 2e 8e 3e

(a) General properties

Aluminium atoms are a little less electropositive than magnesium. They are also smaller, giving rise to stronger metallic bonding. Aluminium atoms

Table 11.5 *the position of aluminium in group 3*

1	2				3	4	5	6	7	0
Li	Be				B	C	N	O	F	Ne
Na	Mg	transition metals			*Al*	Si	P	S	Cl	Ar
K	Ca		Fe	Cu	Zn	Ge			Br	Kr
									I	Xe

ionise through loss of the three outer electrons to achieve inert-gas stability. This metal
(i) has a white lustrous shine,
(ii) conducts electricity and heat well, and is malleable and ductile

> *because of metallic bonding*

(iii) is harder, with a slightly higher melting point and heat of fusion, than magnesium

> *because metallic bonding is stronger than that of magnesium*

(iv) has a higher density than magnesium,

> *because its atoms are smaller than those of magnesium*

but is nevertheless exceptionally light compared to most metals (it is nearly three times lighter than iron).

(b) Chemical reactions

(i) *The apparent stability of aluminium*

Aluminium is found immediately below magnesium in the electrochemical series (see Section 8.5), and would therefore be expected to show similar reactivity. However, magnesium burns in air with a brilliant flame, whereas aluminium kitchen pans are never found to catch fire in the oven. If aluminium metal really did show a similar reactivity to magnesium, it would certainly not be used for aircraft or tube-train bodies.

Aluminium should be very reactive, but generally it is not. This unexpected stability is the result of a tough coating of aluminium oxide which forms spontaneously on the surface when the metal is brought into contact with air (Fig. 11.10). It is as though the metal has been carefully wrapped up in a protective sheet. Thus aluminium articles do not burn in air, nor are they attacked by water, nor easily corroded.

fig 11.10 *the oxide protection of aluminium*

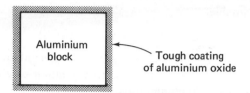

However, when finely powdered and heated strongly, the metal will react with nearly the same vigour as magnesium. For example, finely powdered aluminium dust will explode if sparked in air, and there is a violent reaction when aluminium powder is heated in the presence of dry chlorine gas:

$$4Al(s) + 3O_2(g) \longrightarrow 2Al_2O_3(s)$$
aluminium oxide

$$2Al(s) + 3Cl_2(g) \longrightarrow 2AlCl_3(s)$$
aluminium chloride

If mercury is poured onto a sheet of aluminium, the oxide coating can no longer stick to the surface. A spongy mass of powdery aluminium oxide is seen to grow from the metal as it now reacts spontaneously with oxygen in the air.

(ii) *Reaction with acids*

Sulphuric acid and nitric acid are both oxidising agents which serve only to thicken the protective oxide coating. Therefore aluminium is not readily attacked by either of these acids. The metal will react with the non-oxidising hydrochloric acid:

$$6HCl(aq) + 2Al(s) \longrightarrow 2AlCl_3(aq) + 3H_2(g)$$

(iii) *Reaction with alkalis*

Unlike group 1 and 2 metals, aluminium dissolves in a strongly alkaline solution to form a complex salt and hydrogen gas:

$$2Al(s) + 2NaOH(aq) + 6H_2O(l) \longrightarrow 2NaAl(OH)_4(aq) + 3H_2(g)$$
sodium aluminate

(c) Compounds of aluminium

The aluminium ion is colourless. Compounds of aluminium are therefore white or transparent.

(i) *Oxide and hydroxide—amphoteric nature*

Aluminium oxide is a stable unreactive compound. It can be obtained by heating the hydroxide:

$$2Al(OH)_3(s) \longrightarrow Al_2O_3(s) + 3H_2O(l)$$

Aluminium hydroxide is formed as a gelatinous precipitate when sodium hydroxide solution is added dropwise to a solution of aluminium ions (for example, aqueous aluminium sulphate):

$$Al^{3+}(aq) + 3OH^-(aq) \longrightarrow Al(OH)_3(s)$$
$$\text{ionic gelatinous precipitate}$$
$$\text{of aluminium hydroxide}$$

However, if more sodium hydroxide solution is added, this precipitate dissolves, because aluminium hydroxide reacts with more sodium hydroxide to form the complex salt, sodium aluminate, which is soluble:

$$Al(OH)_3(s) + NaOH(aq) \longrightarrow NaAl(OH)_4(aq)$$
$$\text{sodium aluminate}$$

Here aluminium hydroxide has reacted with an *alkali* to form a *salt*. Thus aluminium hydroxide has behaved as an *acid*. Being a metal hydroxide, it will also behave as a *base*:

$$\underset{\text{(base)}}{Al(OH)_3(s)} + \underset{\text{(acid)}}{3HCl(aq)} \longrightarrow \underset{\text{(salt)}}{AlCl_3(aq)} + \underset{\text{(water)}}{3H_2O(l)}$$

Thus aluminium hydroxide can behave as an acid (it reacts with alkali) or as a base (it reacts with acid). It is said to be *amphoteric*.

Aluminium oxide shows similar *amphoteric* behaviour.

(i) Behaving as a base:

$$\underset{\text{(base)}}{Al_2O_3(s)} + \underset{\text{(acid)}}{6HCl(aq)} \longrightarrow \underset{\text{(salt)}}{2AlCl_3(aq)} + \underset{\text{(water)}}{3H_2O(l)}$$

(ii) Behaving as an acid:

$$\underset{\text{(acid)}}{Al_2O_3(s)} + \underset{\text{(base)}}{3H_2O(l) + 2NaOH(aq)} \longrightarrow \underset{\text{(salt)}}{2NaAl(OH)_4(aq)}$$

(ii) *Polarising effects*

Being a smaller, more highly charged ion than those of groups 1 or 2, Al^{3+} is more polarising (see Section 10.2).

(1) Aluminium chloride

Aluminium chloride readily vaporises on heating. A truly ionic chloride would be non-volatile, but here the polarising aluminium ion has pulled electrons back from the Cl^- ions to give a high degree of covalency.

(2) Hydration of salts

Aluminium compounds show a marked affinity for water, since the polarising Al^{3+} ion strongly attracts the negative ends of water molecules towards it.

Aluminium salts are exceptionally *deliquescent*, forming pools of solution within even an hour of exposure to the air.

(d) Aluminium in everyday life
Aluminium has two major advantages over iron and steel.
(i) It is more resistant to atmospheric corrosion.
(ii) It has a much lower density.
It is therefore widely used in place of steel when weight or corrosion might be a serious problem. Hence its use in aircraft construction, and in the bodywork of many trains. Aluminium does suffer from disadvantages.
(i) It is more costly to produce than steel.
(ii) It cannot be welded easily.
(iii) It corrodes seriously in sea water.
Bodywork of aluminium is easily recognised because the shiny metal surface is usually left exposed. Whereas iron has to be given a protective coat of paint, aluminium has provided its own oxide coating.

Pure aluminium does not possess the same strength as iron, but some aluminium alloys show a strength approaching that of steel. An example is *duralumin*, in which the metal is alloyed with copper and magnesium.

The corrosion resistance of aluminium is often increased by thickening the protective oxide layer. This can be achieved by making an aluminium article the anode in an electrolysis of chromic acid. Oxygen is formed at the anode, which reacts to increase the thickness of the oxide layer. This thickened layer has the added advantage that it will then adsorb dyes to produce, for example, the familiar coloured kitchen pots and pans.

Weight for weight, aluminium conducts electricity 2.7 times better than copper. For this reason, it is used in the overhead cables of the national electricity grid. Perhaps the most familiar use of aluminium is in kitchen metal foil, milk-bottle tops and toothpaste tubes.

11.5 THE TRANSITION METALS

Although we have seen that some metals can float on water, burst into flames in air and even be cut by a knife, the layman would probably say that a metal had to be a heavy, strong substance which clanged when dropped and which could be dented easily; the sort of thing that a bucket or a bridge would be made from. Nearly all such 'typical' metals are transition metals. Iron, copper, silver, gold, nickel and chromium are all transition metals.

(a) Position in the Periodic Table

Table 11.6 shows the position of the transition metals in the Periodic Table.

Table 11.6 *part of the Periodic Table, showing where the transition metals appear*

Group 1	Group 2		Group 3 etc.
Lithium 2, 1	Beryllium 2, 2		Boron 2, 3
Sodium 2, 8, 1	Magnesium 2, 8, 2		Aluminium 2, 8, 3
Potassium 2, 8, 8, 1	Calcium 2, 8, 8, 2	*ten transition metals in which the third energy level fills from 8 electrons to 18 electrons*	Gallium 2, 8, 18, 3

Calcium, in group 2 of the Periodic Table, has the electronic structure 20p 2e 8e 8e 2e. The next element to go into group 3 must have three outermost electrons. However, the next ten elements after calcium show little similarity with aluminium in group 3, but all have very similar properties to each other. They are therefore placed together in one elongated series called 'the transition metals' (transitional between groups 2 and 3).

It is believed that, after calcium, extra electrons of the next ten elements go into the third energy level, filling it from 8 to its maximum of 18. Thus iron, atomic number 26, has the structure 26p 2e 8e 14e 2e, and nickel, atomic number 28, has the structure 28p 2e 8e 16e 2e. Only after zinc, 30p 2e 8e 18e 2e, will electrons continue to fill the fourth energy level, allowing the element gallium, $_{31}$Ga 31p 2e 8e 18e 3e, to be placed in group 3 below aluminium.

Iron and copper are two transition metals which are frequently met in everyday life, and they are used here as typical examples.

(b) General properties of transition metals

The extra ten electrons adding to the third energy level seem to be ineffective at screening the positive nucleus (see Section 10.1); outer electrons are pulled in tightly towards the nucleus. Thus transition metal ions are polarising (see Section 10.2) and their atoms are small, giving rise to strong metallic bonding.

The atoms show stability by losing varying numbers of electrons, even though an inert-gas configuration is rarely left behind. As a consequence, a transition metal frequently has differently charged ions. Thus iron can lose two electrons or three electrons to form Fe^{2+} or Fe^{3+} ions. Copper will lose one or two electrons to form Cu^+ or Cu^{2+} ions.

The transition metals
  (i) are shiny, malleable, ductile solids which are good conductors of heat and electricity

because they are metallically bonded

 (ii) are tough strong metals with high melting points and high heats of fusion

because the metallic bonding is strong

(iii) have high density

because the atoms are small

(c) Chemical reactions of the transition metals
The transition metals
  (i) generally show a low reactivity

because outer electrons are tightly held, so not easily given away

 (ii) have several differently charged ions

because different numbers of electrons can be given away

(iii) have coloured ions

because outer electrons absorb some light energies as they jump within the third energy level

(iv) make good catalysts

by using the extra electrons in the third energy level the metal forms temporary bonds which aid reaction

11.6 IRON—A TYPICAL TRANSITION METAL

(a) General properties
Iron is a shiny metal which exhibits all the general properties of transition metals outlined in Section 11.5.

(b) Ionic charge

Like all metals, iron reacts by giving away electrons. Being a transition metal, it forms differently charged ions:

$$Fe \longrightarrow Fe^{2+} + 2e$$

In this case iron(II) compounds are formed.

$$Fe \longrightarrow Fe^{3+} + 3e$$

In this case iron(III) compounds are formed.

(c) Reaction with air

If iron wool is heated to red heat in a bunsen flame for a short time, a black powdery solid forms on the surface of the strands. This is a mixture of iron(II) oxide and iron(III) oxide with the formula Fe_3O_4. This mixed oxide is called iron(II) diiron(III) oxide.

(d) Rusting

An experiment to examine rusting of iron is shown in Fig. 11.11. This experiment shows the need for both air and moisture before rusting can occur. Analysis shows rust to be hydrated iron(III) oxide, $Fe_2O_3xH_2O$, where the value of x varies.

(e) Reaction with dilute acids

Being above hydrogen in the electrochemical series, iron is more favourable as ions than hydrogen. Hence

$$Fe(s) + 2H^+(aq) \longrightarrow Fe^{2+}(aq) + H_2(g)$$
$$\text{in any acid}$$

For example, we have

$$Fe(s) + 2HCl(aq) \longrightarrow FeCl_2(aq) + H_2(g)$$

(f) Reaction with chlorine

Chlorine is strongly electronegative. It therefore takes three rather than two electrons from each iron atom, to form iron(III) chloride:

$$2Fe(s) + 3Cl_2(g) \longrightarrow 2FeCl_3(s)$$

Hot iron wool, when plunged into a gas jar of chlorine, glows brightly and iron(III) chloride is formed as a fine yellow smoke of tiny particles.

(g) Compounds of iron

The Fe^{2+} ion, and therefore iron(II) compounds, are pale green. The Fe^{3+} ion, and therefore iron(III) compounds, are yellow-brown. Being

fig 11.11 *requirements for rusting*

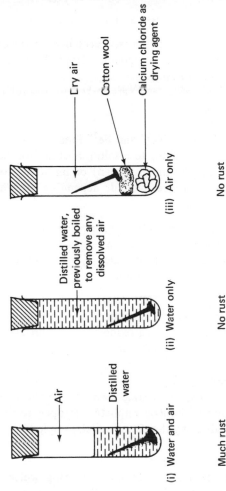

Iron nail in each sealed test tube

Air →

Distilled water →

Distilled water, previously boiled to remove any dissolved air →

Dry air →

Cotton wool →

Calcium chloride as drying agent →

(i) Water and air

(ii) Water only

(iii) Air only

Result after leaving for 1 week:

Much rust

No rust

No rust

small, both ions are polarising. Thus iron(II) and iron(III) hydroxides, carbonates and nitrates decompose to the oxide on heating:

$$FeCO_3(s) \longrightarrow FeO(s) + CO_2(g)$$

$$2Fe(OH)_3(s) \longrightarrow Fe_2O_3(s) + 3H_2O(l)$$

The polarising nature of both ions also accounts for the appreciable degree of hydration in many iron compounds. Iron(III) chloride is highly *deliquescent* (see Section 11.2(c)).

Iron oxides, hydroxides and carbonates are insoluble in water. Other common compounds dissolve well.

(h) Distinguishing between Fe^{2+} and Fe^{3+} ions
If sodium hydroxide is added to a solution of an iron salt in water, insoluble iron(II) hydroxide (muddy green) or insoluble iron(III) hydroxide (yellow-brown) will be precipitated, depending on whether iron(II) or iron(III) ions were present:

$$Fe^{2+}(aq) \quad + \quad 2OH^-(aq) \longrightarrow Fe(OH)_2(s)$$

any iron(II) compound	in sodium hydroxide solution	muddy-green precipitate

$$Fe^{3+}(aq) \quad + \quad 3OH^-(aq) \longrightarrow Fe(OH)_3(s)$$

any iron(III) compound	in sodium hydroxide solution	yellow-brown precipitate

(i) Which oxidation state?
In general the 3+ oxidation state seems to be the more stable. Thus a pale-green solution of iron(II) sulphate left open to the air turns slowly yellow as it is oxidised by the air to iron(III) sulphate.

(j) Iron in everyday life
It is difficult to find a side of modern living which involves no iron. Boats, cars, industrial machinery, bridges, refrigerators, cookers, watches, office blocks all contain iron. Pure iron is surprisingly easy to deform. A bridge made of pure iron would soon stretch and sag into the river. If mixed with a trace of carbon, together with different transition metals, much tougher *steel* is produced (see Section 11.8(b)).

11.7 COPPER—A TYPICAL TRANSITION METAL

(a) General properties
Copper is a pink shiny metal with all the general properties of a transition metal given in Section 11.5.

(b) Ionic charge

Like iron, copper forms ions in two oxidation states:

$$Cu \longrightarrow Cu^+ + 1e$$

In this case copper(I) compounds are formed.

$$Cu \longrightarrow Cu^{2+} + 2e$$

In this case copper(II) compounds are formed.

Copper, being lower in the electrochemical series than iron, is less ready to give away electrons to form these ions. The 2+ state is the more stable. Copper(II) compounds are nearly always obtained when copper metal reacts.

(c) Reaction with oxygen

When a strip of copper is heated strongly by a bunsen flame, a thin black layer of copper(II) oxide forms on the surface:

$$2Cu(s) + O_2(g) \longrightarrow 2CuO(s)$$

Once cooled, this layer may be scraped off to leave pink unreacted copper beneath.

(d) Reaction with the air

Surface copper reacts slowly in air to form a green protective coating of copper(II) carbonate and copper(II) hydroxide. This is the familiar pale-green colour seen on the copper roofs and domes of many large buildings.

(e) Reaction with acids

Copper is positioned below hydrogen in the electrochemical series. Thus, compared with hydrogen, copper is the more favourable as atoms. If copper metal (composed of atoms) is dipped into a dilute acid (containing $H^+(aq)$ ions), copper will remain as atoms. Thus copper does not react with dilute acids.

However, a strongly oxidising acid, such as nitric acid, can oxidise (pull electrons away from) copper atoms to leave copper(II) ions in solution (Fig. 11.12). The essential reaction is

$$Cu(s) \xrightarrow[\substack{\text{nitric acid as an} \\ \text{oxidising agent}}]{\text{electrons removed by}} Cu^{2+}(aq)$$

Blue copper(II) nitrate solution is formed.

(f) Compounds of copper

Copper(II) ions are blue or green, depending upon their environment. For

fig 11.12 *the reaction of nitric acid with copper*

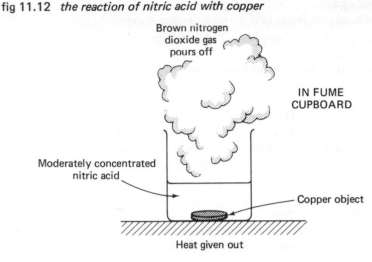

The polarising nature of Cu^{2+} also accounts for the hydration of many copper(II) compounds. For example, blue copper(II) sulphate crystals are hydrated: $CuSO_4.5H_2O$. If heated strongly, the water of hydration is driven away as steam to leave a white solid of $CuSO_4$. Once cooled, addition of water to this *anhydrous* solid brings back both the blue colour and the heat:

example, the hydrated sulphate and nitrate are clear blue, whereas the carbonate and chloride are blue-green. Copper(II) oxide is black because in this compound all energies of visible light are absorbed by electrons jumping between the copper and oxide ions.

Since Cu^{2+} is a polarising ion, copper(II) hydroxide, carbonate and nitrate each decompose to the oxide on heating:

$$Cu(OH)_2(s) \longrightarrow CuO(s) + H_2O(l)$$

$$CuCO_3(s) \longrightarrow CuO(s) + CO_2(g)$$

$$2Cu(NO_3)_2(s) \longrightarrow 2CuO(s) + 4NO_2(g) + O_2(g)$$

The polarising nature of Cu^{2+} also accounts for the hydration of many copper(II) compounds. For example, blue copper(II) sulphate crystals are hydrated: $CuSO_4.5H_2O$. If heated strongly, the water of hydration is driven away as steam to leave a white solid of $CuSO_4$. Once cooled, addition of water to this *anhydrous* solid brings back both the blue colour and the heat:

$$CuSO_4.5H_2O(s) \xrightleftharpoons[\text{heat given out}]{\text{put heat in}} CuSO_4(s) + 5H_2O(l)$$

White anhydrous copper sulphate can be used in testing for water. If a liquid turns it from white to blue, then water must be present.

Copper(II) oxide, hydroxide and carbonate are insoluble in water. Other common copper(II) compounds dissolve readily.

(g) Testing for Cu^{2+} ions
(i) If an aqueous solution of sodium hydroxide is added to a solution containing Cu^{2+}(aq) ions, a pastel blue precipitate of copper(II) hydroxide appears.

(ii) The flame test (see Section 11.1) shows a blue-green coloration when Cu^{2+} ions are present.

(h) Copper(I) compounds
Copper(I) compounds are rare. Copper(I) oxide appears as a red-brown precipitate when copper(II) sulphate solution is reduced:

$$Cu^{2+}(aq) + \underset{\substack{\text{electron given by} \\ \text{a reducing agent} \\ \text{(reducing agents} \\ \text{donate electrons}}}{1e} \longrightarrow Cu^+$$

The details of the reduction are more complicated. First a solution of sodium potassium tartrate, made alkaline by the addition of sodium hydroxide, is added to copper(II) sulphate solution. Tartrate ions surround the Cu^{2+} ions to produce a deep-blue solution known to biologists as 'Fehling's solution'. A reducing agent, for example a little solid glucose, is added and the mixture warmed. The solution turns green, and then a red-brown precipitate of copper(I) oxide appears, which can be filtered off, washed and dried.

(i) Copper in everyday life
In the air, copper slowly forms a protective covering of copper(II) carbonate and hydroxide. Further corrosion is then prevented. However, the expense of the metal means that it is now rarely used as a cladding for roofs and domes of large buildings.

Copper is an easily 'worked' metal; it is particularly malleable and can be beaten into shape easily. Alloyed with tin, *bronze* was one of the first metals used by primitive man, because of its ease of refining and working. *Brass* is an alloy of copper and zinc showing equally good working properties. Being highly ductile, copper is used to produce pipes which can be bent into shape and round corners with ease. The high conductivity of the metal accounts for its extensive use in electrical wiring.

11.8 THE COMMERCIAL MANUFACTURE OF METALS AND COMPOUNDS

(a) Extracting metals from the Earth
Ore is metal-bearing rock. In Table 11.7 the common ores of some metals are shown in relation to their position in the electrochemical series. Silver

Table 11.7 *common metal ores*

	Metal	Ore	Formula
more	K	carnallite	$KCl.MgCl_2.6H_2O$
favourable	Ca	limestone	$CaCO_3$
as ions	Na	rock salt	$NaCl$
↑	Mg	carnallite	$KCl.MgCl_2.6H_2O$
	Al	bauxite	Al_2O_3
	Zn	zinc blende	ZnS
↓	Fe	haematite	Fe_2O_3
more	Cu	copper pyrites	$CuS.FeS$
favourable	Ag	native	Ag
as atoms	Au	native	Au

and gold, being favourable as atoms, are found in the Earth's crust as the free element. They are said to occur *native*, and need only to be physically separated from veins in rock. The principle of 'panning for gold' uses the high density of this metal to keep specks of it at the bottom when crushed rock and water are agitated in a pan. Some copper occurs native, but metals higher in the series are too reactive and are found only as ions in the Earth's crust as the metal oxide, carbonate, sulphate, etc. Metal ions, being positive, must gain electrons to form atoms:

$$M^{x+} + xe \longrightarrow M$$
$$\text{ions} \qquad\qquad \text{atoms}$$

Thus, to obtain the free metal, a metal ore has to be *reduced*. Carbon monoxide and carbon are both relatively cheap reducing agents manufactured from coal. However, although effective in the reduction of iron and zinc ores, stronger reducing agents must be found for metals higher in the electrochemical series. For such metals, the most versatile and controllable reducing agent, or electron supplier, is electricity itself. Metal ions are given electrons by the cathode in the electrolysis of molten oxide (for aluminium) or chloride (for magnesium, sodium, calcium and potassium).

Details of the industrial reduction of sodium, aluminium and iron ores will illustrate the main features involved in extracting metals from the Earth.

(i) *Sodium*

The industrial manufacture of sodium is shown in Fig. 11.13. The electrolyte of sodium chloride is mixed with calcium chloride and calcium fluoride in order to lower the melting point from 800°C to 600°C. Once molten,

fig 11.13 *the industrial manufacture of sodium*

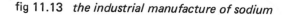

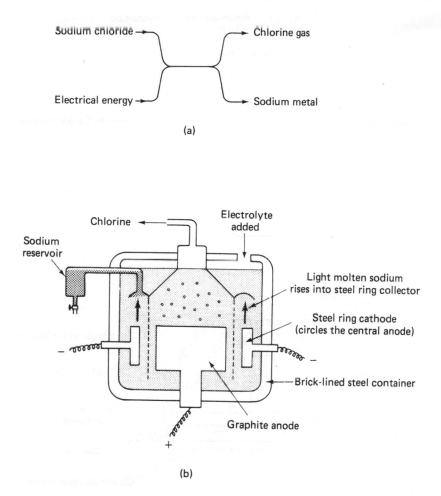

(b)

the electrical resistance within the cell is sufficient to maintain the temperature without external heating.

Cathode reaction

$$Na^+ + e \longrightarrow Na(s)$$

Anode reaction

$$2Cl^- \longrightarrow 2e + Cl_2(g)$$

(ii) *Aluminium*

The industrial manufacture of aluminium is shown in Fig. 11.14.

fig 11.14 *the industrial manufacture of aluminium*

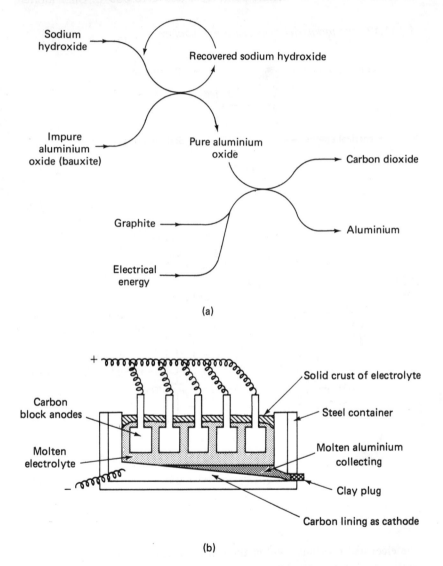

(a)

(b)

Bauxite, hydrated aluminium oxide, contains impurities of iron(III) oxide and sand. Being amphoteric, aluminium oxide reacts with alkali to form the soluble salt, sodium aluminate (see Section 11.4(c)). Thus the powdered ore is dissolved under pressure in hot sodium hydroxide solution which is then drained from the insoluble impurities:

$$Al_2O_3(s) + 2NaOH(aq) + 3H_2O(l) \longrightarrow 2NaAl(OH)_4(aq)$$
$$\text{sodium aluminate}$$

The addition of a little aluminium hydroxide to the solution encourages more to precipitate out. Aluminium hydroxide is said to be *seeded out* from the solution:

$$NaAl(OH)_4(aq) \xrightarrow[\text{with some}]{\text{seeding}} NaOH(aq) + Al(OH)_3(s)$$
$$Al(OH)_3$$

Heating converts the solid hydroxide into pure aluminium oxide which is then reduced by electrolysis:

$$2Al(OH)_3(s) \longrightarrow Al_2O_3(s) + 3H_2O(l)$$

The electrolyte of aluminium oxide is mixed with *cryolite* (rock containing aluminium and sodium fluorides) in order to reduce the melting point from $2000°C$ to $950°C$. Electrical resistance within the cell is sufficient to maintain the temperature during electrolysis.

Cathode reaction

$$Al^{3+} + 3e \longrightarrow Al(s)$$

Anode reaction

$$2O^{2-} \longrightarrow 4e + O_2(g)$$

but then

$$O_2(g) + C(s) \longrightarrow CO_2(g) + Heat$$

The carbon anode blocks are steadily eaten away by reacting with oxygen to produce carbon dioxide, but the heat generated does help to maintain the temperature in the voltameter.

(iii) *Iron*

The industrial manufacture of iron is shown in Fig. 11.15. The highly exothermic reaction forming carbon monoxide raises the temperature at the furnace centre to $1500°C$, and it is here that iron ore is reduced to molten iron. The heat also decomposes limestone to calcium oxide

$$CaCO_3(s) \longrightarrow CaO(s) + CO_2(g)$$
$$\text{limestone}$$

which forms molten calcium silicate slag with the sandy impurities:

$$CaO(s) + SiO_2(s) \longrightarrow CaSiO_3(l)$$
$$\text{sand} \qquad \text{slag}$$

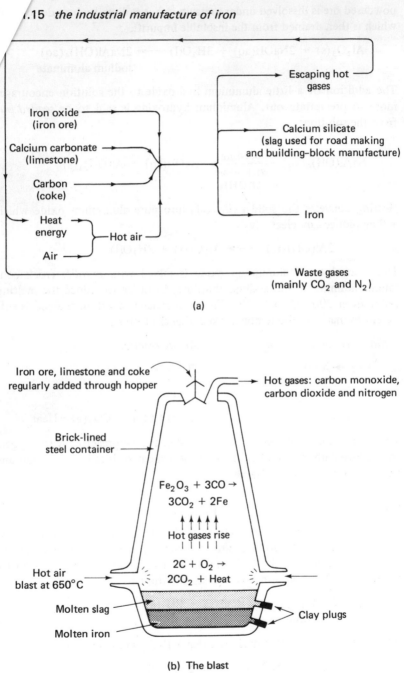

(a)

Iron ore, limestone and coke regularly added through hopper

Hot gases: carbon monoxide, carbon dioxide and nitrogen

Brick-lined steel container

$Fe_2O_3 + 3CO \rightarrow 3CO_2 + 2Fe$

Hot gases rise

$2C + O_2 \rightarrow 2CO_2 + Heat$

Hot air blast at 650°C

Molten slag

Molten iron

Clay plugs

(b) The blast furnace

This is tapped off from the surface of the molten iron.

Iron from the blast furnace contains about 6% carbon. It is cast into solid ingots of *pig iron*.

(b) Steel manufacture

Pure iron is highly malleable and ductile, but not especially strong. As its percentage carbon content increases, so the strength increases, but at the same time the metal becomes less malleable and ductile. Pig iron from the blast furnace containing about 6% carbon is an exceptionally hard brittle metal. In the Bessemer process for manufacturing steel, hot air is blown through molten pig iron to remove carbon and other impurities as gaseous oxides. In the Linz Donawitz process, pure oxygen is blown through. By then adding precise small quantities of carbon to the pure molten iron, steel of the required strength and malleability is obtained. The carbon content of steel varies from 0.1% in mild steels to 1.5% in hard steels. Other transition metals are also added in order to modify the properties of the metal. For example, the addition of chromium (18%) and nickel 8%) produces a stainless steel.

(c) Sodium hydroxide

Sodium hydroxide, known commercially as caustic soda, is one of the most important industrial chemicals (see Section 11.2(d)). It is obtained in bulk by electrolysing aqueous sodium chloride in a mercury cathode voltameter (Fig. 11.16).

Mercury flows steadily along the base of the voltameter, and sodium is deposited in preference to hydrogen, combining with the mercury to form *sodium amalgam* (see Section 8.10).

Cathode reaction

$$Na^+(aq) + e \rightarrow Na(s)$$

Anode reaction

$$2Cl^-(aq) \rightarrow Cl_2(g) + 2e$$

Sodium, Na(s), combines
with mercury to form
sodium amalgam.

Sodium amalgam is pumped into a second tank where water is added to produce aqueous sodium hydroxide and hydrogen gas:

$$2Na(s) + 2H_2O(l) \longrightarrow 2NaOH(aq) + H_2(g)$$

Mercury is recycled through the cell. Chlorine and hydrogen are useful by-products of this process.

fig 11.16 *the industrial manufacture of sodium hydroxide*

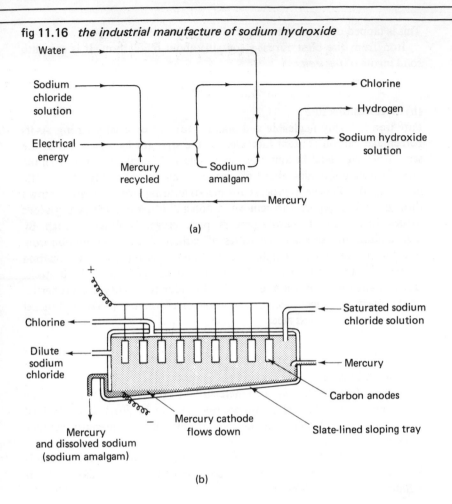

(a)

(b)

(d) Cement and concrete

Pulverised limestone, clay, sand and coal are moved steadily along a rotating steel tunnel against a counter-blast of air. Reaction between the coal and air generates the high temperature needed to fuse the ingredients into lumps of complex salts containing calcium, silicon and aluminium. This is powdered into *cement*, which sets into a solid rocklike mass on the addition of water. A mixture of cement with gravel is *concrete*, and if this is set around steel rods exceptionally strong *reinforced concrete* is obtained.

CHAPTER 12

THE NON-METALS

The non-metals are found to the right of the Periodic Table, as shown in Table 12.1. They bond covalently with each other, usually to form individual small molecules: for example as diatomic molecules

O_2, N_2, Cl_2, F_2, Br_2, I_2

or polyatomic molecules

P_4, S_8.

Table 12.1 *the non-metals*

1	2				3	4	5	6	7	0
Li	Bo				*B*	*C*	*N*	*O*	*F*	*Ne*
Na	Mg	transition metals			Al	*Si*	*P*	*S*	*Cl*	*Ar*
K	Ca		Fe	Cu Zn					*Br*	*Kr*
									I	*Xe*

In a few non-metallic elements, notably carbon and silicon, bonding continues from one atom to the next in a giant structure (see Section 3.7).

When non-metals react with metals, electron transfer takes place to form ionically bonded compounds.

When a non-metal reacts with another non-metal, electrons are shared to form a covalently bonded compound.

12.1 GROUP 7, THE HALOGENS

The *halogens* are shown in *italics* in Table 12.2, and are listed below.

Table 12.2 *the halogens*

1	2					3	4	5	6	7	0
Li	Be					B	C	N	O	*F*	Ne
Na	Mg	transition metals				Al	Si	P	S	*Cl*	Ar
K	Ca		Fe	Cu	Zn	Ge				*Br*	Kr
										I	Xe

fluorine, F   9p   2e   7e
chlorine, Cl   17p   2e   8e   7e
bromine, Br   35p   2e   8e   18e   7e
iodine, I

The members of this reactive group of non-metals show many similarities.

(a) General properties

The halogen atoms are small, and only one electron short of the stable inert-gas electron arrangement. They therefore show a strong tendency to react with other atoms by gaining one electron. Fluorine, the smallest, most electronegative element, shows the greatest attraction for electrons, and iodine the least. The atoms combine with each other to form diatomic molecules, and the weak secondary bonding between these molecules is greatest between molecules of iodine and least between those of fluorine.

(i) *State*

Being composed of individual molecules, the halogens are volatile. Because the secondary bonding between the molecules increases down the group, fluorine and chlorine are gases at ordinary temperatures, bromine is a liquid and iodine a volatile solid.

(ii) *Smell*

The halogens all have a similar choking smell.

(iii) *Colour*

The colour becomes progressively darker down the group:
    fluorine   pale yellow
    chlorine   yellow–green
    bromine   red–brown
    iodine   black (purple as vapour)

(iv) *Solubility in water*

Iodine and bromine dissolve slightly in water to give brown solutions. Chlorine dissolves to give a pale yellow-green solution, but also reacts partially with water molecules to produce halogen acids (see below). Fluorine does not dissolve, but attacks water vigorously (see below).

(b) Chemical reactions

(i) Reaction with metals

Most of the halogens react with most metals by taking electrons from them. Fluorine is the most reactive (greatest attraction for electrons) and iodine the least reactive. The products are metal *halides*. A few examples are given below.

(1) Iron and chlorine

When hot iron wool is lowered into a gas jar of chlorine, it splutters and glows red hot as a brown smoke of iron(III) chloride particles is produced:

$$2Fe(s) + 3Cl_2(g) \longrightarrow 2FeCl_3(s)$$
$$\text{iron(III) chloride}$$

(2) Zinc and bromine

Similarly, if zinc wool is lowered into an atmosphere of bromine vapour, it sparks and gives an off-white smoke of zinc bromide particles:

$$Zn(s) + Br_2(g) \longrightarrow ZnBr_2(s)$$
$$\text{zinc bromide}$$

(3) Aluminium and iodine

A mixture of powdered aluminium and iodine shows similar reactivity. A few drops of water are added to start the reaction off. Within a minute, purple fumes and sparks pour from the mixture. The fumes consist of excess iodine vapourising away, leaving behind solid aluminium iodide:

$$2Al(s) + 3I_2(s) \longrightarrow 2AlI_3(s)$$
$$\text{aluminium iodide}$$

In each of these examples, the halogen has taken electrons from the metal to form negative halide ions. Metal halides are generally ionic crystalline solids which, with the exception of silver and lead halides, are soluble in water.

(ii) *Reaction with non-metals*

Most of the halogens react directly with most non-metals to form covalently bonded halides (Fig. 12.1).

fig 12.1 *reacting a non-metal with chlorine*

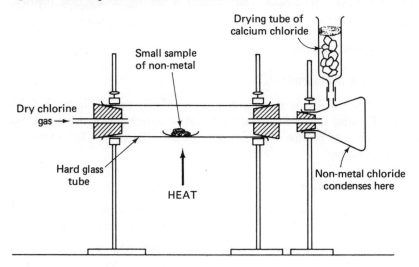

(1) Phosphorus and chlorine

With the apparatus shown, phosphorus reacts with chlorine to produce a mixture of phosphorus pentachloride and phosphorus trichloride:

$$2P(s) + 5Cl_2(g) \longrightarrow 2PCl_5(s)$$
covalent phosphorus
pentachloride

$$2P(s) + 3Cl_2(g) \longrightarrow 2PCl_3(l)$$
covalent phosphorus
trichloride

(2) Silicon and chlorine

Similarly, silicon reacts with chlorine to form liquid silicon tetrachloride:

$$Si(s) + 2Cl_2(g) \longrightarrow SiCl_4(l)$$
covalent silicon
tetrachloride

These reactions, once started, are highly exothermic. The apparatus must be kept dry because covalent chlorides (with the exception of tetrachloromethane, CCl_4) are attacked by water to form hydrogen chloride gas. For example, phosphorus pentachloride hisses on contact with water forming phosphoric(V) acid and hydrogen chloride gas. For this reason, covalent chlorides have to be kept tightly sealed from the atmosphere in order to prevent them from reacting with moisture in the air.

All covalent chlorides are composed of individual molecules. All are therefore volatile solids, liquids or even gases at ordinary temperatures.

(iii) *Reaction with hydrogen*

The halogens react with hydrogen to form covalent hydrogen halide gases:

$$H_2(g) + F_2(g) \xrightarrow[\text{mixing}]{\text{explodes on}} 2HF(g)$$

$$H_2(g) + Cl_2(g) \xrightarrow[\substack{\text{ultra-violet} \\ \text{light}}]{\text{explodes in}} 2HCl(g)$$

$$H_2(g) + Br_2(g) \xrightarrow[\text{slowly}]{\text{reacts}} 2HBr(g)$$

$$H_2(g) + I_2(g) \xrightleftharpoons[\text{reaction}]{\text{reversible}} 2HI(g)$$

In the case of reaction with bromine and iodine, the halogen must be vaporised by heating before reaction can start.

The resulting hydrogen halides are gases which dissolve well in water to form acidic solutions. Hydrogen chloride is more conveniently prepared in the laboratory from common salt and concentrated sulphuric acid (Fig. 12.2):

$$H_2SO_4(l) + NaCl(s) \longrightarrow NaHSO_4(s) + HCl(g)$$

(iv) *Properties of hydrogen chloride gas*

In the experiment shown in Fig. 12.3, the syringe of hydrogen chloride becomes filled completely with water, suggesting that the gas has completely dissolved. The indicator turns red, showing the resulting solution to be acidic.

Hydrogen chloride is exceptionally soluble in water. A sample of water will dissolve over 450 times its own volume of the gas. The resulting solution is hydrochloric acid:

fig 12.2 *laboratory preparation of hydrogen chloride*

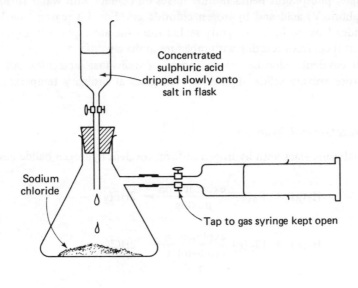

Concentrated
sulphuric acid
dripped slowly onto
salt in flask

Sodium
chloride

Tap to gas syringe kept open

$$HCl(g) \ + \ H_2O(l) \longrightarrow H_3O^+(aq) \ + \ Cl^-(aq)$$

or

$$HCl(g) \ + \ Aq \longrightarrow H^+(aq) \ + \ Cl^-(aq)$$

fig 12.3 *the reaction of hydrogen chloride with water*

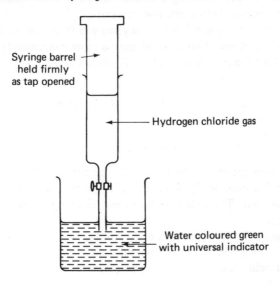

Syringe barrel
held firmly
as tap opened

Hydrogen chloride gas

Water coloured green
with universal indicator

Being completely ionised in solution, this is a strong acid (see Section 9.10).

When a drop of concentrated aqueous ammonia solution is held in hydrogen chloride gas, a dense white smoke of tiny particles of ammonium chloride forms:

$$NH_3(g) + HCl(g) \longrightarrow NH_4Cl(s)$$
$$\text{ionic ammonium chloride}$$

This is often used as a quick test for the gas.

(v) Reaction with water

Fluorine attacks water violently to form oxygen gas and the highly corrosive hydrogen fluoride:

$$2F_2(g) + 2H_2O(l) \longrightarrow 4HF(l) + O_2(g)$$

Chlorine dissolves slightly in water, and the dissolved molecules react reversibly to form hydrochloric acid and chloric(I) acid:

$$Cl_2(g) + H_2O(l) \longrightarrow HCl(aq) + HClO(aq)$$

| | hydrochloric acid | chloric(I) acid |

(c) Testing for halogens

(i) General test

If aqueous solutions containing chlorine, bromine or iodine are shaken with tetrachloromethane (a dense liquid which does not mix with water), the halogen will dissolve in the latter to give a characteristic colour.

Yellow-green shows the presence of chlorine.

Red-brown shows the presence of bromine.

Purple shows the presence of iodine.

(ii) Special test for chlorine

Being electronegative, chlorine removes electrons from (i.e. it oxidises) many coloured dyes. Colour results from outer electrons jumping. By removing these electrons, a colour will no longer be given. Thus chlorine *bleaches* moistened indicator paper.

(iii) Special test for iodine

Aqueous solutions of iodine give an intense blue-black colour with starch.

(d) Testing for halide ions
Silver chloride, bromide and iodide are insoluble in water, and have characteristic colours.

Dilute nitric acid is first added to the test solution in order to destroy any carbonate ions since these, like halide ions, also give a precipitate with silver ions. Aqueous silver nitrate is then added.

A white precipitate shows chloride to be present (the precipitate is silver chloride).

A cream precipitate shows bromide to be present (the precipitate is silver bromide).

A yellow precipitate shows iodide to be present (the precipitate is silver iodide).

$$Ag^+(aq) \ + \ Halide^-(aq) \ \longrightarrow \ AgHalide(s)$$

in silver nitrate	in halide solution under test	solid silver halide precipitate

(e) Halogen-halide interaction
Electrons are readily exchanged between halogens and halide ions. This is illustrated in the experiment shown in Fig. 12.4. In this experiment, *iodide ions* have become *iodine molecules*. The *reason* for this can be written as

chlorine
bromine   *electronegativity*
iodine   *increasing*

Of these three elements, chlorine gains electrons most readily forming Cl^- ions. If chlorine is mixed with iodide ions
  (i) iodide loses electrons to become iodine (more favourable as molecules), and
 (ii) chlorine gains electrons to become chloride (more favourable as ions)

(electrons transferred)

$$Cl_2(aq) \ + \ 2I^-(aq) \ \longrightarrow \ 2Cl^-(aq) + I_2(aq)$$

The results of similar experiments are summarised in Table 12.3.

(f) The halogens in everyday life
Fluorine is used in the preparation of polytetrafluoroethene (PTFE or 'Teflon') (see Section 14.3). Hydrogen fluoride is used to etch patterns onto ornamental glass.

Chlorine is widely used in bleaches, and many town water supplies

fig 12.4 *the reaction of iodide ions with chlorine molecules*

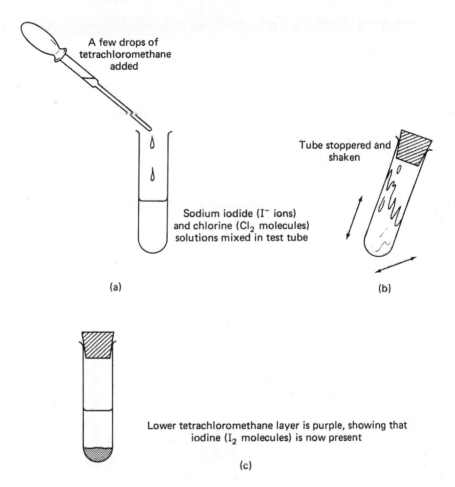

A few drops of
tetrachloromethane
added

Tube stoppered and
shaken

Sodium iodide (I⁻ ions)
and chlorine (Cl₂ molecules)
solutions mixed in test tube

(a)   (b)

Lower tetrachloromethane layer is purple, showing that
iodine (I₂ molecules) is now present

(c)

and swimming baths are chlorinated in order to oxidise and thus destroy bacteria. The gas is also used to manufacture chloroethene (old name, vinylchloride) from which polyvinylchloride, PVC, is obtained (see Section 14.3).

A coating of fine particles of silver bromide suspended in gelatine is the basis of light-sensitive photographic paper. When light strikes silver bromide, Ag^+ ions are reduced to a dark area of metallic silver, and in this way a photographic negative is produced.

A solution of iodine in alcohol is 'tincture of iodine', used to oxidise and kill bacteria in cuts on the skin.

Table 12.3

(a) *Summary of some halogen–halide interactions*

	Halide solution	*Halogen solution*	*Colour of tetrachloromethane after reaction*	*Therefore halogen present after reaction*	*Reaction or not?*
(1)	sodium bromide	chlorine	red–brown	bromine	yes
(2)	sodium iodide	bromine	purple	iodine	yes
(3)	sodium bromide	iodine	purple	iodine	no
(4)	sodium iodide	chlorine	purple	iodine	yes
(5)	sodium chloride	bromine	red–brown	bromine	no

(b) *Equations and conclusions from these interactions*

(1) $2Br^-(aq) + Cl_2(aq) \rightarrow 2Cl^-(aq) + Br_2(aq)$
Therefore chlorine is more favourable as ions than bromine

(2) $2I^-(aq) + Br_2(aq) \rightarrow 2Br^-(aq) + I_2(aq)$
Therefore bromine is more favourable as ions than iodine.

(3) $2Br^-(aq) + I_2(aq) \rightarrow$ no reaction
Therefore iodine is not more favourable as ions than bromine.

(4) $2I^-(aq) + Cl_2(aq) \rightarrow 2Cl^-(aq) + I_2(aq)$
Therefore chlorine is more favourable as ions than iodine.

(5) $2Cl^-(aq) + Br_2(aq) \rightarrow$ no reaction
Therefore bromine is not more favourable as ions than chlorine.

12.2 THE AIR

The air is a mixture of a number of non-metallic elements and compounds.

(a) What is in air?

In the apparatus shown in Fig. 12.5, precisely 100 cm³ of trapped air is

fig 12.5 *determining the oxygen content of the air*

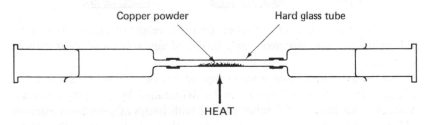

Copper powder   Hard glass tube

HEAT

pushed from one syringe to the other and back again, giving up its oxygen to the heated copper as it passes:

$$2Cu(s) + O_2(g) \longrightarrow 2Cu(s)$$

The pink copper reacts to form black copper(II) oxide. The volume of trapped air can be seen to decrease. Once there is no further volume change, heating is stopped and the volume of gas, when all pushed into one syringe, is seen now to be only about 80 cm³. Thus twenty-hundredths (20/100) or one-fifth of the air is seen to be oxygen. Small samples of the gas remaining in the syringe can be tested as follows.

(i) A lighted splint is extinguished by the gas. The remaining gas therefore contains little or no oxygen (the splint would have continued to burn brightly), and little or no hydrogen (the gas would have given a pop).

(ii) A few drops of lime water shaken with a small sample of the gas does not go milky. The remaining gas therefore contains little or no carbon dioxide.

By such processes of elimination, the gas is concluded to be largely nitrogen. Air is one-fifth oxygen and four-fifths nitrogen by volume. However, the air does contain traces of other gases. If a *large volume* of air is bubbled for some time through lime water, a milkiness does eventually appear. Air does contain a little carbon dioxide.

The quantity of carbon dioxide in air can be determined experimentally by slowly pumping a known volume of dry air through two 'U' tubes packed with pellets of potassium hydroxide (Fig. 12.6). Carbon dioxide is completely absorbed by potassium hydroxide:

$$2KOH(s) + CO_2(g) \longrightarrow K_2CO_3(s) + H_2O(l)$$

fig 12.6 *determining the carbon dioxide content of the air*

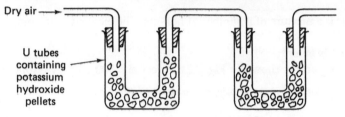

The increase in weight of the tubes gives the weight of carbon dioxide in this volume of air. Approximately 0.03% of any volume of air is carbon dioxide.

Air usually contains some water vapour, but the proportion depends on climate. The amount of water can be determined by pumping a known volume of air through 'U' tubes packed with lumps of anhydrous calcium chloride. This solid takes up water to form the hydrated salt, $CaCl_2.6H_2O$. The increase in the weight of the tubes now gives the weight of water in this volume of air. Air contains between 0% (dry air over a desert region) and 6% (humid air over tropical jungle) water vapour by volume.

Nitrogen is sometimes prepared in the laboratory by passing it over
(i) heated copper to remove the oxygen,
(ii) anhydrous calcium chloride to remove the water vapour, and
(iii) potassium hydroxide pellets to remove the carbon dioxide.
However, the gas obtained in this way is denser than pure nitrogen because of the presence of tiny amounts of the dense inert gases, neon, argon, krypton and xenon. Discovery of the very existence of the inert gases came from this observation. The inert gases occupy 0.95% by volume of air.

Both oxygen and nitrogen are important industrial gases, and their cheapest source is the air. Commercially these gases are separated from the air by liquefaction followed by fractional distillation (see Section 12.10(a)).

12.3 OXYGEN AND THE OXIDES

Oxygen occurs in group 6 of the Periodic Table (Table 12.4) and has electron arrangement
oxygen, O 8p 2e 6e

(a) General properties
Oxygen is found in the air as diatomic molecules, O_2, with a double co-valent bond. The second bond is weak, consequently oxygen molecules are reactive. Oxygen is highly electronegative, attracting towards it another

Table 12.4 *the position of oxygen in group 6*

1	2				3	4	5	6	7	0
Li	Be				B	C	N	*O*	F	Ne
Na	Mg	transition metals			Al	Si	P	S	Cl	Ar
K	Ca		Fe	Cu Zn		Ge			Br	Kr
									I	Xe

two electrons to gain the inert-gas configuration. Only fluorine is more electronegative than oxygen. Oxygen is a colourless gas with no smell.

(b) Chemical reactions

Oxygen reacts with all metals except those at the very bottom of the electrochemical series (see Table 8.5). The metal atoms are oxidised (electrons removed) to positive ions whilst oxygen is itself reduced to the oxide ion, O^{2-}. A metal can be heated in tongs until ignited or red hot, and then plunged into a gas jar of oxygen.

With magnesium, reaction is highly exothermic, the metal burning instantaneously with a brilliant white light:

$$2Mg(s) + O_2(g) \longrightarrow 2MgO(s)$$
$$\text{magnesium oxide}$$

With iron wool, the metal glows red hot and sparks fly from its surface:

$$4Fe(s) + 3O_2(g) \longrightarrow 2Fe_2O_3(s)$$
$$\text{iron(III) oxide}$$

Oxygen reacts with most non-metals to form covalent molecules. Experimentally the reactions can be carried out as described above for metals.

With carbon, the heated solid bursts into brilliant flame when it comes into contact with the pure oxygen:

$$C(s) + O_2(g) \longrightarrow CO_2(g)$$
$$\text{carbon dioxide}$$

With sulphur, an intense blue flame is observed:

$$S(s) + O_2(g) \longrightarrow SO_2(g)$$
$$\text{sulphur dioxide}$$

Nitrogen and the halogens combine with oxygen only at very high temperatures, and the inert gases do not react with oxygen at all.

(c) Properties of oxides.

The metallic oxides are bases, reacting with acid to form a salt and water:

$$CuO(s) + H_2SO_4(aq) \longrightarrow CuSO_4(aq) + H_2O(l)$$

base   acid   salt   water

Metal oxides, being ionic, are non-volatile solids.

Non-metallic oxides are acidic. If the combustion products of the non-metals from the above experiments are shaken with water, the addition of a few drops of universal indicator solution will show an acid coloration:

$$SO_2(g) + H_2O(l) \longrightarrow H_2SO_3(aq)$$
sulphurous acid

$$CO_2(g) + H_2O(l) \longrightarrow H_2CO_3(aq)$$
carbonic acid

Non-metallic oxides, being molecular, are volatile solids, liquids or gases. The one exception is silicon dioxide, which is a solid with a high melting point and with giant covalent structure (see Section 3.7).

Elements near the metal–non-metal boundary of the Periodic Table have *amphoteric* oxides. These react in some circumstances as acids and in others as bases. Aluminium oxide is amphoteric (see Section 11.4(c)).

(i) Reaction as a base:

$$Al_2O_3(s) + 6HCl(aq) \longrightarrow 2AlCl_3(aq) + 3H_2O(l)$$
base   acid   salt

(ii) Reaction as an acid:

$$Al_2O_3(s) + 2NaOH(aq) + 3H_2O(l) \longrightarrow 2NaAl(OH)_4(aq)$$
acid   base   salt

Hydrogen oxide, water, is considered separately in Section 12.9.

(d) Laboratory preparation of oxygen

(i) By adding hydrogen peroxide solution to powdered manganese(IV) oxide (Fig. 12.7). Manganese(IV) oxide catalyses the decomposition of hydrogen peroxide:

$$2H_2O_2(l) \longrightarrow 2H_2O(l) + O_2(g)$$

(ii) By heating various higher oxides and oxy-salts, for example potassium manganate(VII) (Fig. 12.8). The equation for this reaction is complex.

(e) Testing for oxygen

Because substances burn so much better in pure undiluted oxygen than in the air, oxygen gas will relight a glowing splint.

fig 12.7 *the laboratory preparation of oxygen, method (i)*

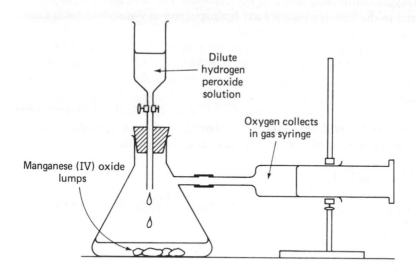

Dilute hydrogen peroxide solution

Oxygen collects in gas syringe

Manganese (IV) oxide lumps

fig 12.8 *the laboratory preparation of oxygen, method (ii)*

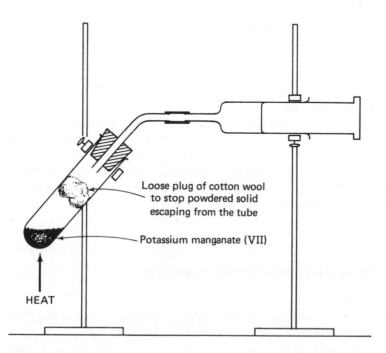

Loose plug of cotton wool to stop powdered solid escaping from the tube

Potassium manganate (VII)

HEAT

(f) Oxygen in everyday life

Oxygen compressed into cast iron cylinders is used in breathing apparatus, and in the high-temperature oxy-hydrogen and oxy-acetylene torch flames used for welding and cutting steel.

The pure gas is used to oxidise away impurities in the Linz Donawitz process for manufacturing steel (see Section 11.8(b)). Liquid oxygen, along with liquid hydrogen, comprises the propellant of the heavy space rockets.

Apart from its obvious use in breathing, oxygen in the air is responsible for a wealth of diverse effects including the fading of dyes, the rotting of fruit, the rusting of iron and the hardening of oil paints.

12.4 SULPHUR

Sulphur occurs in group 6 of the Periodic Table (Table 12.5) and its electron arrangement is

sulphur, S   16p 2e 8e 6e

Table 12.5   *the position of sulphur in group 6*

1	2					3	4	5	6	7	0
Li	Be					B	C	N	O	F	Ne
Na	Mg	transition metals				Al	Si	P	S	Cl	Ar
K	Ca		Fe	Cu	Zn	Ge				Br	Kr
										I	Xe

(a) General properties

Sulphur occurs naturally as the free element in large deposits in the USA and Sicily. Although less electronegative than oxygen, sulphur atoms still readily achieve the inert-gas stability by gaining two electrons. However, because its outer energy level can hold up to eighteen electrons (see Fig. 2.7), sulphur atoms will also share four and six electrons. For example, the oxides of sulphur are sulphur dioxide

$$O \diagup\!\!\!= S =\!\!\!\diagdown O$$

(sulphur sharing four electrons) and sulphur trioxide

$$\begin{array}{c} O \\ \| \\ O \diagup\!\!\!= S =\!\!\!\diagdown O \end{array}$$

(sulphur sharing six electrons). These oxides have a stable existence even though here sulphur has not achieved an inert-gas structure. The reason for this is beyond the scope of this book.

Sulphur is found as S_8 molecules, in which each sulphur shares an electron with its two neighbours in a zig-zag ring.

The S_8 molecule

Sulphur is a yellow solid. It is obtained from chemical suppliers either as solid sticks (roll sulphur) or as a fluffy powder (flowers of sulphur), but under the microscope both types can be seen to consist of the same compact crystalline shape, shown below.

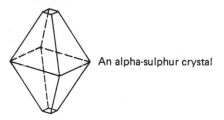

An alpha-sulphur crystal

This crystalline form, alpha-sulphur (α-sulphur) is stable at room temperature. In contrast, if sulphur is dissolved in hot toluene (toluene is a highly flammable organic liquid which must not be heated by a naked flame) and the solution allowed to cool, sulphur crystals now appear as fine yellow needles, shown below.

A beta-sulphur crystal

This different crystalline arrangement of the S_8 molecules is called beta-sulphur (β-sulphur). It is the stable form above 96.5°C. If left to cool further, the needles slowly break up into smaller alpha-sulphur crystals.

Since it can exist in more than one crystalline shape, sulphur is said to be *polymorphic*. Alpha-sulphur and beta-sulphur are both *polymorphs* of sulphur.

If heated gently in a test tube, sulphur melts, at about 113°C, to a runny yellow liquid. However, as more heat is put in, the liquid is transformed into a thick viscous syrup which hardly flows at all. Still more heating creates a runny liquid once more which vaporises at 445°C.

This unusual behaviour has been accounted for in terms of the sulphur molecule's structure.

	heat	heat	
solid sulphur	→ *runny liquid*	→ *very viscous liquid*	┐
compact S_8 rings	compact S_8 rings move easily around and past one another	heat energy breaks rings into sulphur chains which get tangled around one another	↓

vapour ←— heat —— *runny liquid* ←— heat ┘
S_2 molecules — heat energy breaks chains into very compact S_2 molecules

If viscous liquid sulphur is rapidly cooled by pouring into cold water, a 'gooey' chewing-gum-like *plastic sulphur* is obtained in which the chains intertwine around one another in a shapeless non-crystalline mass. Within five minutes, this is a brittle solid mass of alpha-sulphur once more.

(b) Chemical reactions

Being a solid, a small volume of sulphur contains many more atoms than the same volume of oxygen gas. Although less electronegative than oxygen, sulphur shows vigorous reactions with many metals and non-metals as a result of this greater concentration of atoms.

Like oxygen, sulphur removes electrons from metal atoms: sulphur oxidises metals. For example sulphur reacts exothermically with aluminium powder:

$$2Al(s) + 3S(s) \longrightarrow Al_2S_3(s)$$
aluminium sulphide

A similar reaction occurs with powdered zinc and sulphur:

$$Zn(s) + S(s) \longrightarrow ZnS(s)$$
zinc sulphide

With non-metals, covalent sulphides are formed. The reaction of sulphur with oxygen is the most common example. Sulphur burns in air with a blue flame to form a colourless gas with a choking smell. The gas is sulphur dioxide:

$$S(s) + O_2(g) \longrightarrow SO_2(g)$$

(c) Oxides and oxoacids of sulphur

(i) *Sulphur dioxide*

If a little water is swirled around a gas jar of sulphur dioxide, the fumes are seen to dissolve quickly. Universal indicator is turned red by this solution, showing it to be acidic. Sulphurous acid has been formed:

$$SO_2(g) + H_2O(l) \longrightarrow H_2SO_3(aq)$$
$$\text{sulphurous acid}$$

In solution, sulphur dioxide is readily oxidised to sulphur trioxide (see below). Thus orange potassium dichromate(VI), which is an oxidising agent, reacts to form green chromium(III) ions. Paper soaked in potassium dichromate(VI) solution is therefore turned from orange to green by sulphur dioxide, and this is used as a convenient test for the gas.

(ii) *Sulphurous acid*

Sulphurous acid is diprotic (see Section 9.8). Complete neutralisation with sodium hydroxide yields sodium sulphite:

$$2NaOH(aq) + H_2SO_3(aq) \longrightarrow Na_2SO_3(aq) + 2H_2O(l)$$
$$\text{sodium}$$
$$\text{sulphite}$$

If only half this quantity of sodium hydroxide is used, then sodium hydrogen sulphite is obtained:

$$NaOH(aq) + H_2SO_3(aq) \longrightarrow NaHSO_3(aq) + H_2O(l)$$
$$\text{sodium}$$
$$\text{hydrogen}$$
$$\text{sulphite}$$

This acid salt is also formed when excess sulphur dioxide is bubbled through sodium hydroxide solution:

$$NaOH(aq) + SO_2(g) \longrightarrow NaHSO_3(aq)$$

Test for the sulphite ion, SO_3^{2-}

All sulphites react spontaneously with dilute acids to produce sulphur dioxide, which can be identified by testing the gas with moist potassium dichromate(VI) paper (see above):

$$CaSO_3(s) + 2HCl(aq) \longrightarrow CaCl_2(aq) + SO_2(g) + H_2O(l)$$

(iii) Sulphur trioxide and sulphuric acid

Sulphuric acid is prepared industrially by the contact process, in which sulphur dioxide is reacted with oxygen in the air to form sulphur trioxide:

$$2SO_2 + O_2(g) \longrightarrow 2SO_3(g)$$

The reaction is catalysed by platinum metal or by vanadium(V) oxide. This process is considered in detail in Section 12.10(c). Sulphur trioxide reacts extremely violently with water to form sulphuric acid:

$$SO_3(g) + H_2O(l) \longrightarrow H_2SO_4(l)$$

Sulphuric acid is
  (i) a strong acid, being almost completely ionised in solution,
 (ii) a mild oxidising agent, and
(iii) a powerful dehydrating agent.

The dehydrating nature of the acid is illustrated by the following two experiments.

(1) If a little concentrated sulphuric acid is added to a beaker of sugar in a fume cupboard, a hot spongy mass of carbon rises like a science fiction monster from the reaction as water is removed from this carbohydrate.

(2) If concentrated sulphuric acid is left in an open beaker, its volume apparently *increases* with time. This results from the absorption of water vapour from the air.

Being a diprotic acid, both normal and acid salts can exist:

$$2NaOH(aq) + H_2SO_4(aq) \longrightarrow Na_2SO_4(aq) + 2H_2O(l)$$
sodium
sulphate

$$NaOH(aq) + H_2SO_4(aq) \longrightarrow NaHSO_4(aq) + H_2O(l)$$
sodium
hydrogen
sulphate

Most sulphates are soluble in water. Lead sulphate and barium sulphate are insoluble.

Test for the sulphate ion, SO_4^{2-}

If barium chloride solution is added to a solution containing sulphate ions, a thick white precipitate of barium sulphate forms:

$$Ba^{2+}(aq) + SO_4^{2-}(aq) \longrightarrow BaSO_4(s)$$

Dilute hydrochloric acid is first added to destroy any sulphite ions, SO_3^{2-}, or carbonate ions, $CO_3{}^{2-}$, since these would also give a precipitate with $Ba^{2+}(aq)$ ions.

(d) Sulphur in everyday life

The main use of sulphur is to supply the large quantity of sulphur dioxide required by the contact process for the manufacture of sulphuric acid. The acid is itself used for the manufacture of many plastics and synthetic fibres, dyes, drugs, detergents, steel, explosives and fertilisers.

Flowers of sulphur are used in certain skin creams as a mild antiseptic, and as a fungicide for plants.

Rubber is naturally very elastic and too squashy for most purposes. By heating natural rubber with sulphur, the long-chain rubber molecules become bonded to one another by short sulphur chains, and are thus held more rigidly. In this way rubber is *vulcanised* to make it tougher. This process was invented in 1838 by Charles Goodyear.

Sulphur is one of the constituents of gunpowder.

12.5 NITROGEN

Nitrogen occurs in group 5 of the Periodic Table (see Table 12.6) and has the following electron arrangement

nitrogen, N   7p 2e 5e

Table 12.6   *the position of nitrogen in group 5*

1	2				3	4	5	6	7	0
Li	Be				B	C	*N*	O	F	Ne
Na	Mg	transition metals			Al	Si	P	S	Cl	Ar
K	Ca		Fe	Cu	Zn	Ge			Br	Kr
									I	Xe

(a) General properties

Nitrogen is found in the air as diatomic molecules, N_2, with a triple covalent bond. Nitrogen is less electronegative than oxygen, and its triple bond binds the atoms strongly. Nitrogen is therefore an unreactive element. Nitrogen is a colourless gas with no smell.

(b) Chemical reactions

Nitrogen reacts readily with very few metals. Only magnesium and lithium burn in the gas to form ionic nitrides:

$$3Mg(s) + N_2(g) \longrightarrow Mg_3N_2(s)$$
magnesium nitride
(contains Mg^{2+} and N^{3-} ions)

$$6Li(s) + N_2(g) \longrightarrow 2Li_3N(s)$$
lithium nitride
(contains Li^+ and N^{3-} ions)

Few non-metals react directly with nitrogen. Oxygen and hydrogen do react under extreme conditions to form nitrogen oxides and ammonia respectively.

(c) Ammonia

This is prepared industrially by the Haber process, in which nitrogen and hydrogen gases combine exothermically in the presence of an iron catalyst:

$$N_2(g) + 3H_2(g) \rightleftharpoons 2NH_3(g)$$

Details of the process are given in Section 12.10(b).

Ammonia is a colourless gas with a pungent smell. It is exceptionally soluble in water. Dissolved ammonia molecules react partially with water to form *ammonium ions*, NH_4^+, and hydroxide ions, OH^-:

$$NH_3(aq) + H_2O(l) \rightleftharpoons NH_4^+(aq) + OH^-(aq)$$

A solution of ammonia is therefore weakly alkaline, turning universal indicator blue. It can be neutralised by acids to produce ammonium salts:

$$NH_3(aq) + HCl(aq) \longrightarrow NH_4Cl(aq)$$
ammonium chloride
(comprising NH_4^+ and Cl^- ions)

$$2NH_3(aq) + H_2SO_4(aq) \longrightarrow (NH_4)_2SO_4(aq)$$
ammonium sulphate
(comprising NH_4^+ and SO_4^{2-} ions)

Ammonia and hydrogen chloride gases react spontaneously when mixed to form a smoke of fine particles of ammonium chloride:

$$NH_3(g) + HCl(g) \longrightarrow NH_4Cl(s)$$
tiny particles of solid
ionic ammonium chloride

This is used as a quick test for the presence of ammonia gas. The gas is also easily recognised by its smell, and by the fact that it turns moist universal indicator paper blue.

(d) Oxides and oxoacids of nitrogen

Rain in a thunderstorm is extremely dilute nitric acid. This is because there is sufficient energy in a lightning flash for nitrogen and oxygen in the air to react forming acidic oxides of nitrogen. The most common oxides of nitrogen are the monoxide, NO, and the dioxide, NO_2.

(i) *Nitrogen monoxide, NO*

Nitrogen monoxide is a colourless gas. When mixed with the air it reacts instantly with oxygen to form brown nitrogen dioxide, NO_2:

$$2NO(g) + O_2(g) \longrightarrow 2NO_2(g)$$

Nitrogen monoxide is obtained when ammonia reacts with oxygen in the presence of a platinum catalyst. This is the first stage in producing nitric acid (see Section 12.10(b)).

(ii) *Nitrogen dioxide, NO_2*

Nitrogen dioxide is a brown poisonous gas. Being a non-metallic oxide it is acidic, turning moist universal indicator paper red. It reacts with water to form nitric and nitrous acids:

$$2NO_2(g) + H_2O(l) \longrightarrow \underset{\substack{\text{nitric} \\ \text{acid}}}{HNO_3(aq)} + \underset{\substack{\text{nitrous} \\ \text{acid}}}{HNO_2(aq)}$$

Laboratory preparation of nitrogen dioxide

Nitrates, except for those of sodium, potassium and ammonia, give off nitrogen dioxide when heated (Fig. 12.9). Oxygen is given off too:

$$2Pb(NO_3)_2(s) \longrightarrow 2PbO(s) + 4NO_2(g) + O_2(g)$$

Oxygen collects in gas syringe (b). Tube (a) collects a green liquid with formula N_2O_4. If this *dinitrogen tetraoxide* is allowed to warm up it vaporises into pale yellow dinitrogen tetraoxide gas. Heating this gas still further causes it to darken, turning more brown as the molecules *dissociate* into nitrogen dioxide, NO_2:

$$\underset{\substack{\text{yellow} \\ \text{gas}}}{N_2O_4(g)} \underset{\text{cool}}{\overset{\text{warm}}{\rightleftharpoons}} \underset{\substack{\text{brown} \\ \text{gas}}}{2NO_2(g)}$$

(iii) *Nitric acid*

The industrial production of nitric acid is detailed in Section 12.10(b). It is a strong acid, being completely ionised in water:

fig 12.9 *the laboratory preparation of nitrogen dioxide*

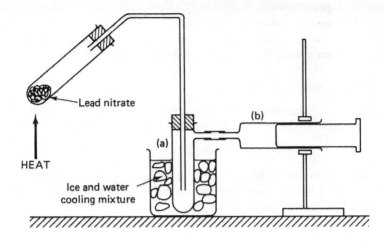

$$HNO_3(aq) \rightleftharpoons H^+(aq) + NO_3^-(aq)$$

The acid is also a good oxidising agent. For example, Section 11.7(e) showed how the acid oxidised copper metal to copper(II) ions.

The acid is neutralised by bases to form nitrate salts, all of which are highly soluble in water:

$$2HNO_3(aq) + ZnO(s) \longrightarrow Zn(NO_3)_2(aq) + H_2O(l)$$
$$\text{zinc nitrate}$$

(e) Nitrogen in everyday life

(i) *Elemental nitrogen*

In air, nitrogen behaves as a *diluent*, diluting the effect of oxygen. Without this dilution, wood, plastic, cloth and paper would have to be treated as dangerous, highly flammable chemicals, a cigarette would burn through in a matter of seconds and we would soon die from the resulting increase in our own metabolic rate. Nitrogen is obtained in vast quantities from the fractional distillation of liquid air for use in the production of ammonia (see Section 12.10(a)).

(ii) *Ammonia*

Ammonia is used mainly in fertilisers, replacing the considerable amount of nitrogen taken from the soil by plants. It is added as ammonium phosphate(V), ammonium sulphate or even injected directly into the soil at $-50°C$ as liquid ammonia.

Because of its mildly alkaline nature, ammonia attacks grease to form water-soluble detergents (see .Section 14.2) and destroys bacteria. It is therefore used in many household cleaners. Much ammonia is prepared specifically for the manufacture of nitric acid.

(iii) *Nitric acid and nitrates*

Much nitric acid is neutralised with ammonia to produce ammonium nitrate fertiliser. Other nitrates, particularly naturally occurring potassium nitrate, are similarly used as nitrogen-supplying fertilisers. Nitric acid is also used in the production of nitro explosives such as gun-cotton and TNT, in addition to dyes and industrial solvents.

fig 12.10 *the nitrogen cycle*

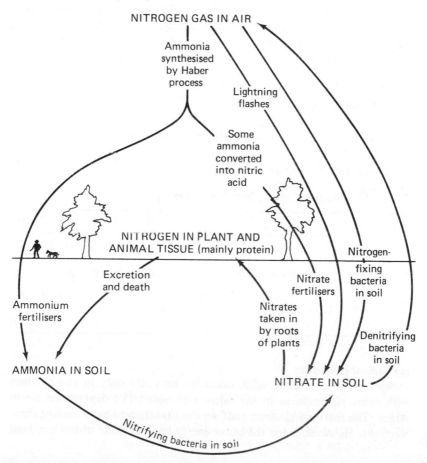

is beyond the scope of this book. Thus the chlorides of phosphorus have formulae PCl_3 (phosphorus sharing three electrons) and PCl_5 (phosphorus sharing five electrons).

(iv) *The nitrogen cycle*

Nitrogen in the plants and animals we eat is held in the chemicals of our bodies and in the chemicals of our excretions. The need for fertilisers arises because very little of our sewage is pumped back onto the fields of corn, wheat, grass, etc., and because our bodies are not buried on farm-land. Nitrogen is taken from the fields for us to eat, but put back elsewhere when we have finished with it. Nitrogen in the air is trapped or *fixed* into nitrogen-containing compounds for adding to the soil by several means.

(i) *Lightning*: nitrogen oxides are formed which dissolve to produce nitric acid.

(ii) *Nitrogen-fixing bacteria*: these live in soil and the roots of some plants, including soya. They convert nitrogen in the air to nitrates.

(iii) *The Haber process*: this converts nitrogen in the air to ammonia.

At the same time, *denitrifying* bacteria in the soil put nitrogen from decaying organic matter back into the air as nitrogen gas. The *nitrogen cycle* (Fig. 12.10) shows these interrelationships.

12.6 **PHOSPHORUS**

Phosphorus is a group 5 element (Table 12.7) whose electron arrangement is

phosphorus, P   15p 2e 8e 5e

Table 12.7   *the position of phosphorus in group 5*

1	2				3	4	5	6	7	0
Li	Be				B	C	N	O	F	Ne
Na	Mg	transition metals			Al	Si	*P*	S	Cl	Ar
K	Ca		Fe	Cu	Zn	Ge			Br	Kr
									I	Xe

(a) General properties
Phosphorus is a reactive solid, occurring naturally only in combination with other elements, as in the calcium phosphate(V) deposits of North Africa. The inert-gas electron configuration is achieved by attracting *three* electrons. However, since the outer energy level of these atoms can hold

up to eighteen electrons, *five* electrons are also found to be shared, even though now the inert-gas structure is not attained. An explanation of this
The element exists as P_4 molecules held loosely to one another in a molecular structure. Two different crystalline states are commonly found. Like sulphur, the element is *polymorphic*. *White phosphorus* is obtained when phosphorus vapour is condensed. It is a very pale-yellow solid, usually obtained in sticks. It reacts violently and spontaneously with air (see below) and for this reason is kept under water. White phosphorus is an unstable polymorph, gradually changing to the stable crystalline form called *violet phosphorus*. However, this change is too slow to be detected at room temperature. Violet phosphorus is less reactive, reacting with air only when sparked or heated.

(b) Chemical reactions

(i) *With air*

White phosphorus begins smoking, then bursts into violent reaction with oxygen in the air, splattering solid white phosphorus(III) and phosphorus(V) oxides in all directions:

$$4P(s) + 3O_2(g) \longrightarrow 2P_2O_3(s)$$
phosphorus(III) oxide

$$4P(s) + 5O_2(g) \longrightarrow 2P_2O_5(s)$$
phosphorus(V) oxide

Initially, reaction occurs only on the surface exposed to air, giving out an eyrie greenish light. This *phosphorescence* explains the element's name.
Violet phosphorus reacts only on heating in air.
Being oxides of a non-metal, both phosphorus(III) and phosphorus(V) oxides react with water to produce acidic solutions.

(ii) *With chlorine*

If a stream of dry chlorine is passed over white phosphorus in the apparatus of Fig. 12.1, reaction begins spontaneously. With violet phosphorus, initial heating is required. A mixture of phosphorus trichloride (a liquid) and phosphorus pentachloride (a white solid) is collected. The apparatus has to be kept dry because both chlorides are attacked by moisture in the air (see Section 12.1(b)):

$$2P(s) + 3Cl_2(g) \longrightarrow 2PCl_3(l)$$
phosphorus trichloride

$$2P(s) + 5Cl_2(g) \longrightarrow 2PCl_5(s)$$
phosphorus pentachloride

(c) Phosphorus in everyday life
Phosphorus is another element essential to life. Phosphates(V) are used in vast quantities as fertilisers in order to supply the phosphorus essential for plant growth. Calcium phosphate(V), found in large deposits in Africa, the USA and Russia, or from the slag left from the blast furnace, is often treated with concentrated sulphuric acid to produce a mixture of calcium phosphate and calcium sulphate. This *superphosphate* fertiliser is more soluble and hence more easily absorbed by plants.

Dilute solutions of phosphoric(V) acid are used in the food industry to give a tart acid taste (for example, in Coca-Cola). Phosphorus itself is used in the production of matches and incendiary bombs.

12.7 CARBON AND SILICON

Carbon and silicon are both found in group 4 (Table 12.8), and their electron arrangements are
    carbon, C   6p 2e 4e
    silicon, Si   14p 2e 8e 4e

Table 12.8   *the positions of carbon and silicon in group 4*

1	2				3	4	5	6	7	0
Li	Be				B	*C*	N	O	F	Ne
Na	Mg	transition metals			Al	*Si*	P	S	Cl	Ar
K	Ca		Fe	Cu	Zn	Ge			Br	Kr
									I	Xe

(a) General properties and structure
Both elements gain stability by sharing four electrons.

(i) *Carbon*

Carbon is polymorphic, being found in the two different crystalline forms, *diamond* and *graphite.*

In *diamond*, each carbon atom shares electrons with four others.

The central carbon atom is seen to have achieved the stable arrangement of eight outer electrons. Each of the surrounding carbon atoms shares electrons with more carbon atoms in a *giant covalent* array. The bonds are positioned tetrahedrally to build up the familiar diamond shape, shown in Fig. 12.11. Each carbon atom is rigidly held in place by four others. To push it out of position would require the distortion of many strong and precisely positioned carbon–carbon bonds. Diamond is therefore one of

fig 12.11 *diamond*

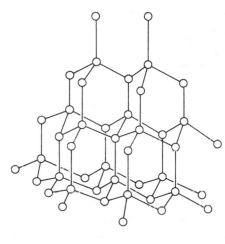

O = Carbon atom

the hardest substances known. The crystals also have an exceptionally high refractive index. This means that light is bent appreciably as it passes in and out of the crystal. By *cleaving* diamond (i.e. by breaking it along the naturally occurring planes of carbon atoms) so that many differently angled faces are exposed, a brilliantly sparkling gemstone can be obtained.

Calculations show that diamond is an unstable polymorph of carbon, slowly changing into the stable polymorph, graphite. At room temperature the change is undetectable, and the World will have ended long before diamonds have become lumps of soot.

Each carbon atom in *graphite* is bonded to only three others in a giant flat array (Fig. 12.12). Only three electrons are involved in sharing. The fact that graphite is a good electrical conductor suggests that the fourth electron is spread out or *delocalised* over the entire plane of carbon atoms. The giant planes are held only loosely to one another by secondary bonding. These planes can slip and slide across one another easily, giving rise to the slippery feel of graphite.

fig 12.12 *graphite*

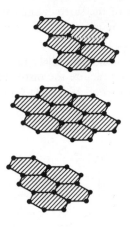

(ii) *Silicon*

Silicon is found in only one form, as a glistening black non-conducting solid with the diamond-type structure.

(b) Occurrence

Coal is compressed decayed vegetation from prehistoric forests. By heating it in the absence of air, many useful organic compounds are driven off as vapour (see Section 13.6). The dark-grey spongy solid left behind, called *coke*, is more-or-less pure carbon. This is used as an excellent slow-burning fuel, and in the reduction of iron ore in the blast furnace (see Section 11.8). In some areas, extreme pressure from movements in the Earth's crust have forced coal into the less-stable polymorph of carbon, and here diamond deposits are found.

Silicon is not found as the free element, but it occurs in vast amounts as silicon dioxide in quartz, sand, onyx and opal, or in silicates. It is second only to oxygen in elemental abundance in the Earth's crust.

(c) Oxides and oxoacids

Carbon burns steadily in air to form carbon dioxide:

$$C(s) + O_2(g) \longrightarrow CO_2(g) \qquad \Delta H_r = -394 \text{ kJ mol}^{-1}$$

The reaction is highly exothermic, making it a good fuel. In a limited supply of air, poisonous carbon monoxide is formed:

$$C(s) + \tfrac{1}{2}O_2(g) \longrightarrow CO(g)$$

Carbon monoxide is often prepared by passing carbon dioxide over heated carbon (Fig. 12.13):

fig 12.13 *the laboratory preparation of carbon monoxide*

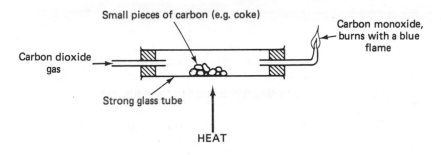

$$CO_2(g) + C(s) \longrightarrow 2CO(g)$$

Carbon monoxide is used industrially as a fuel gas (see Section 12.10(d)). It is a colourless gas which burns with a steady blue flame to give carbon dioxide:

$$2CO(g) + O_2(g) \longrightarrow 2CO_2(g) \qquad \Delta H_r = -220 \text{ kJ mol}^{-1}$$

Carbon dioxide is a colourless gas comprising individual CO_2 molecules. It dissolves slightly in water to form carbonic acid (non-metallic oxides are acidic):

$$CO_2(g) + H_2O(l) \longrightarrow H_2CO_3(aq)$$

This is an exceptionally weak acid (see Section 9.10). Being acidic, carbon dioxide reacts with alkalis to form carbonates. On contact with potassium hydroxide, the gas is completely absorbed:

$$CO_2(g) + 2KOH(s) \longrightarrow K_2CO_3(s) + H_2O(l)$$

When shaken with calcium hydroxide solution (lime water), carbon dioxide gives the familiar milky precipitate of calcium carbonate used as a test for this gas:

$$CO_2(g) + Ca(OH)_2(aq) \longrightarrow CaCO_3(s) + H_2O(l)$$
$$\text{milky ppt}$$

With the exception of sodium and potassium, the metal carbonates are insoluble solids which give off carbon dioxide gas when heated:

$$CaCO_3(s) \longrightarrow CaO(s) + CO_2(g)$$

All carbonates react with dilute acids to produce carbon dioxide, itself detected using lime water. This is used as a test for the carbonate ion, CO_3^{2-}:

$$MgCO_3(s) + H_2SO_4(aq) \longrightarrow MgSO_4(aq) + CO_2(g) + H_2O(l)$$

Silicon reacts violently with oxygen to form silicon dioxide, but intense heating is needed to initiate reaction:

$$Si(s) + O_2(g) \longrightarrow SiO_2(s) \qquad \Delta H_r = -910 \text{ kJ mol}^{-1}$$

Silicon dioxide is a non-volatile solid. It has a giant covalent structure in which each silicon atom is linked to four others via oxygen. Being an acidic oxide, it reacts with alkalis and bases to form silicates:

$$SiO_2(s) + 2NaOH(aq) \longrightarrow Na_2SiO_3(aq) + H_2O(l)$$
$$\text{sodium}$$
$$\text{silicate}$$

$$SiO_2(s) + CaO(s) \longrightarrow CaSiO_3(s)$$
$$\text{calcium}$$
$$\text{silicate}$$

The latter reaction is used in the blast furnace to remove sandy impurities as a molten silicate slag (see Section 11.8).

(d) Carbon and silicon in everyday life
Because of its hardness, powdered chips of diamond are coated onto the surface of drilling bits of oil rigs. A single chip provides the hard-wearing point of a record-playing stylus, and diamond cutters are used to cut glass and to shape stone. The high refraction of diamond gemstones has already been mentioned.

Graphite is used as the electrodes in many industrial electrolytic processes (see Section 11.8). Its slippery nature as its flat planes slide across one another explains its use as a high-temperature lubricant where conventional oil might burn or oxidise. Mixed with clay it also makes the so-called 'lead' of pencils.

Coke is used directly as an industrial fuel, in addition to preparing fuel gases containing carbon monoxide. It is also responsible for the reduction of iron ore to iron in the blast furnace. It is the presence of a trace of carbon which gives steel its strength, and the very existence of life depends upon the stability of long chains of carbon atoms (see Chapter 13).

12.8 HYDROGEN

Hydrogen has the simple electron arrangement
hydrogen, H   1p 1e

(a) Position in the Periodic Table

(i) *In group 7?*

Hydrogen achieves the inert-gas stability of helium by gaining one electron to form the hydride ion, H^-. For example, burning sodium metal in hydrogen gas produces this ion:

$$2Na(s) + H_2(g) \longrightarrow 2NaH(s)$$
sodium hydride
(contains Na^+ and H^- ions)

Comparison with the reaction with chlorine

$$2Na(s) + Cl_2(g) \longrightarrow 2NaCl(s)$$
sodium chloride
(contains Na^+ and Cl^- ions)

suggests that hydrogen should be placed in group 7 with the halogens.

(ii) *In group 1?*

In the presence of water, hydrogen also exists without any outer electrons as $H^+(aq)$. Aqueous solutions of acids contain this ion (see Section 9.3). Hydrogen might therefore be placed in group 1 where singly charged positive ions are readily formed.

The problem of where to put hydrogen in a Periodic Table is often solved by not putting it in at all. This non-committal approach is adopted here.

(b) The preparation of hydrogen

(i) *By electrolysis*

Hydrogen is given off at the cathode in the electrolysis of dilute aqueous acids (see Section 8.9(c)).

(ii) *By displacing hydrogen from acids*

A metal above hydrogen in the electrochemical series will displace hydrogen from dilute acids (see Section 9.4(a)):

$$Zn(s) + 2HCl(aq) \longrightarrow ZnCl_2(aq) + H_2(g)$$

(iii) *By displacing hydrogen from water*

Reactive metals will spontaneously react with water to give hydrogen:

$$Ca(s) + 2H_2O(l) \longrightarrow Ca(OH)_2(aq) + H_2(g)$$

The gas used to be obtained commercially by passing steam over heated iron:

$$3Fe(s) + 4H_2O(g) \rightleftharpoons Fe_3O_4(s) + 4H_2(g)$$

(iv) *From oil*

Today, hydrogen for industry is most commonly obtained as a by-product of the *cracking* of crude oil fractions (see Section 13.6).

(c) General properties

Hydrogen is a diatomic gas, H_2. It is colourless and has no smell. As a result of its exceptionally low density (it is fifteen times lighter than air), no hydrogen is found in the air.

(d) Chemical reactions

(i) *With metals*

Hydrogen reacts only with the most reactive metals to form ionic *hydrides* containing the H^- ion:

$$2Na(s) + H_2(g) \longrightarrow 2NaH(s)$$
$$\text{sodium hydride}$$

(ii) *With non-metals*

Hydrogen reacts directly with a few non-metals, notably the halogens (see Section 12.1(b))

$$H_2(g) + F_2(g) \longrightarrow 2HF(g)$$

and most important of all with oxygen

$$2H_2(g) + O_2(g) \xrightarrow{\text{explodes on}}_{\text{sparking}} 2H_2O(l)$$

Mixtures of hydrogen and oxygen explode when sparked, forming water vapour. For this reason, flames should be kept well away from hydrogen-generating apparatus. In order to test for hydrogen, a small test tube of the gas should be collected and taken to a flame some distance away. A pop shows hydrogen to be present.

(iii) *With metal oxides*

Hydrogen is more reactive than elements below it in the electrochemical series. If hydrogen gas is passed over heated copper(II) oxide, Cu^{2+} ions are reduced by hydrogen to Cu atoms (Fig. 12.14). Black copper(II)

fig 12.14   *the reaction of hydrogen with copper(II) oxide*

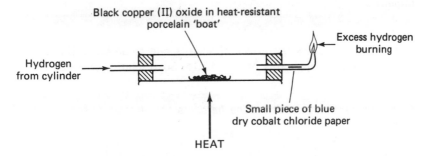

oxide becomes pink copper metal. The cobalt chloride paper turns pink, showing water to have been produced:

$$CuO(s) + H_2(g) \longrightarrow Cu(s) + H_2O(l)$$

(e) Hydrogen in everyday life
The high-pressure treatment of coal with hydrogen produces substitute petrol. This process is rapidly becoming economic as the cost of crude oil rises. Hydrogen is already used in quantity in the commercial production of ammonia (see Section 12.10(b)) and in the hydrogenation of fats from which margarine is manufactured.

Liquid hydrogen together with liquid oxygen comprise the primary fuel of heavy space rockets.

12.9 WATER

This oxide of hydrogen is essential to all life.

(a) Properties of water
Water is a polar molecule in which the oxygen atom holds a slight negative charge, with the hydrogen atoms slightly positive (see Section 3.8(b)). When water freezes into ice, secondary *hydrogen bonds* hold the molecules together in a diamond-type structure. The bulkiness of this arrangement means that the molecules in ice are actually less compact than in water, and it is for this reason that ice forms on the top rather than the bottom of a pond in winter.

Pure water conducts electricity only slightly because of the minimal ionisation:

$$H_2O(l) \rightleftharpoons H^+(aq) + OH^-(aq)$$

Water is neutral because for every $H^+(aq)$ causing acidity there is an $OH^-(aq)$ ion causing alkalinity to balance out the effect. Most water conducts electricity appreciably because of the presence of dissolved ionic impurities.

(b) Hard water

Water that neutralises the effect of soap is said to be *hard*. This can be studied using the apparatus of Fig. 12.15. Soap solution is added to the

fig 12.15 *apparatus to determine hardness of water using soap solution*

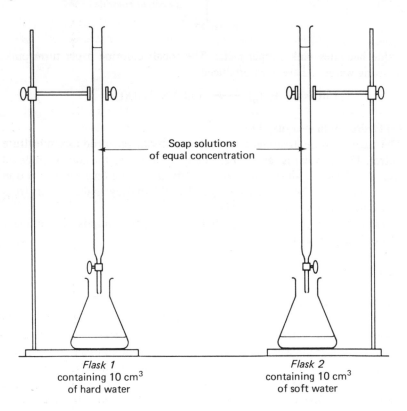

Flask 1
containing 10 cm^3
of hard water

Flask 2
containing 10 cm^3
of soft water

water samples from the burettes, and the flasks are vigorously swirled. A large volume of soap is needed to create a lather with hard water (flask 1) whereas very little is needed to create a lather with the soft water (flask 2).

(i) *Temporary hardness*

Water collected from chalky ground is hard because it contains dissolved calcium ions:

$$CaCO_3(s) \; + \; H_2O(l) \; + \; CO_2(g) \; \longrightarrow \; Ca(HCO_3)_2(aq)$$

solid chalk   rain water   calcium hydrogen
carbonate solution
(contains Ca^{2+} and
HCO_3^- ions)

Soap is sodium stearate, $Na^+ C_{18}H_{35}O_2^-$. Calcium ions from hard water and stearate ions from the soap combine to form a gritty *scum* of calcium stearate:

$$Ca^{2+}(aq) \; + \; 2C_{18}H_{35}O_2^-(aq) \; \longrightarrow \; Ca(C_{18}H_{35}O_2)_2(s)$$

in hard   in soap   solid ionic scum of
water   calcium stearate

Only once all the calcium ions have combined in this way can more soap remain unaffected.

Calcium hydrogen carbonate causes *temporary* hardness because it is quite unstable. If its solution is heated or evaporated, insoluble calcium carbonate precipitates out so that calcium ions are no longer in solution:

$$Ca(HCO_3)_2(aq) \; \longrightarrow \; CaCO_3(s) \; + \; H_2O(l) \; + \; CO_2(g)$$

The 'fur' left in water pipes and kettles in hard-water areas is no more than precipitated calcium carbonate. These articles may be 'de-furred' simply by soaking in a dilute acid such as vinegar.

In limestone caves stalactites and stalagmites form where drops of hard water have evaporated to leave behind solid calcium carbonate.

(ii) *Permanent hardness*

Many rocks contain magnesium or calcium salts which are sparingly soluble in water but are stable to heat. Thus rain water percolating through, for example, dolomite rock (a mixture of magnesium and calcium carbonates) or gypsum (calcium sulphate) will contain some Mg^{2+} or Ca^{2+} ions. These will form scum with soap as described above, but now simple boiling will not precipitate out the rogue ions. Such *permanent* hardness can be removed only by *deionising* or *distilling* the water.

(c) Deionised water

Ions can be removed from water by passing it through a deionising column. This is a tube packed with two *deionising resins*, one to remove cations and the other to remove anions. The first is composed of a giant covalent lattice such as Bakelite (see Section 14.2) with many H^+ ions held loosely to its surface. Positive ions in the water exchange with these ions on the lattice (Fig. 12.16).

fig 12.16 *removal of positive ions from water by a deionising resin*

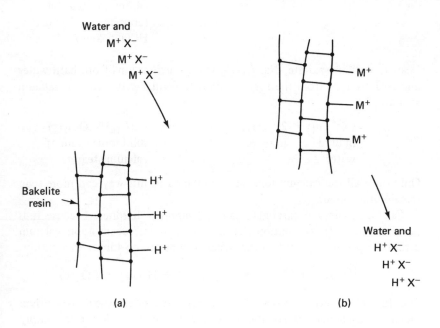

(a)   (b)

fig 12.17 *removal of remaining ions from water by a deionising resin*

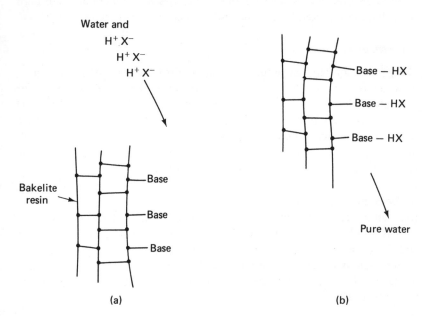

(a)   (b)

The second deionising resin consists of a similar lattice, but this time with bases loosely bonded to its surface. These react with the acids now present (H^+X^-) holding them to the lattice (Fig. 12.17).
Deionised water may still contain non-ionic impurities.

(d) Distilled water
Many ions, along with any non-volatile covalent compounds, can be removed from water by distillation (see Section 4.2(c)).
Distilled water may still contain volatile impurities which can drift across with the water vapour.

(e) Action of soaps and detergents
See Section 14.2

(f) Test for water
Hydrated cobalt chloride, $CoCl_2.6H_2O$, is pink. When paper soaked in a solution of this salt is dried out it turns sky blue. The addition of a drop of water to the paper regenerates the hydrated pink colour. Thus water turns blue cobalt chloride paper pink.

In a similar way, water will restore the blue colour to anhydrous copper(II) sulphate (see Section 11.7(f)).

12.10 THE COMMERCIAL PRODUCTION OF NON-METALS AND THEIR COMPOUNDS

(a) The separation of oxygen and nitrogen from air
Nitrogen and oxygen are obtained in bulk for industry by the fractional distillation of liquid air. Air is liquefied by the following stages (see Fig. 12.18):
  (i) Carbon dioxide and water vapour are frozen out by cooling air to below $-80°C$.
 (ii) The remaining gases are cooled and compressed to 200 atmospheres.
(iii) The compressed air is allowed to escape through a nozzle into an

fig 12.18 *flow diagram of the liquefaction of air*

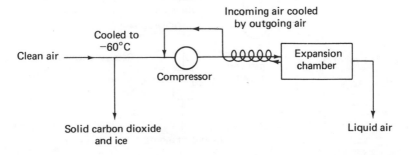

expansion chamber. Sudden expansion makes the gas colder (feel the air escaping from a car or cycle tyre when the valve is opened).

(iv) Repeated recycling through compression and sudden expansion progressively decreases the temperature until, just below −180°C, the air liquefies.

The resulting mixture of liquid nitrogen and oxygen is allowed to boil at the base of a fractionating column (see Section 4.2(c)). Nitrogen gas (boiling point −196°C) is collected from the top of the column, and oxygen (boiling point −183°C) from the bottom.

(b) Nitrogen compounds

(i) *Ammonia*

The bulk of nitrogen separated from air is used by the Haber process to prepare ammonia, NH_3. The process uses the reversible reaction between nitrogen and hydrogen gases in the presence of a finely divided iron catalyst:

$$N_2(g) + 3H_2(g) \rightleftharpoons 2NH_3(g) \qquad \Delta H_r = -92 \text{ kJ mol}^{-1}$$

The application of Le Chatelier's principle to the choice of a moderate running temperature (500°C) and a high pressure (250 atmospheres) was considered in Section 6.4.

The Haber process has the following stages (see Fig. 12.19):

(i) Nitrogen, from the fractional distillation of liquid air, and hydrogen, from the cracking of crude oil, are mixed in the ratio 1:3 by volume.

(ii) The mixed gases are compressed to 250 atmospheres. High pressure improves both the yield and rate of production of ammonia.

(iii) The gases pass through a heat exchanger where their temperature is raised to 500°C by hot exhaust gases from the reaction chamber.

fig 12.19 *flow diagram of the Haber process*

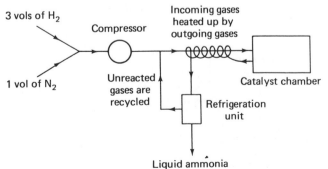

3 vols of H_2

Compressor

Incoming gases heated up by outgoing gases

1 vol of N_2

Unreacted gases are recycled

Catalyst chamber

Refrigeration unit

Liquid ammonia

(iv) In the reaction chamber, ammonia is formed on the large catalytic surface of finely divided. iron. Approximately 7% conversion to ammonia is achieved.

(v) The gases are cooled to $-40°C$. Liquid ammonia is run off and the remaining unreacted nitrogen and hydrogen are recycled.

(ii) *Nitric acid*

Nitric acid is prepared by the catalytic oxidation of ammonia in the following stages (see Fig. 12.20):

(i) Ammonia is mixed with a large excess of air (1:10 by volume).

(ii) After slight compression, the mixture is passed through layers of fine rhodium-platinum gauze catalyst:

$$4NH_3(g) + 5O_2(g) \longrightarrow 4NO(g) + 6H_2O(l) \qquad \Delta H_r = -1170 \text{ kJ mol}^{-1}$$

The highly exothermic reaction maintains the catalyst temperature at $750°C$ without the need for additional heating.

(iii) The gases are allowed to expand into cooling chambers. Here nitrogen monoxide reacts with more of the excess air to form nitrogen dioxide:

$$2NO(g) + O_2(g) \longrightarrow 2NO_2(g)$$

(iv) The gases now pass up a series of towers against a steady trickle of water. The excess air, nitrogen dioxide and water react to produce aqueous nitric acid:

$$4NO_2(g) + O_2(g) + 2H_2O(l) \longrightarrow 4HNO_3(l)$$

fig 12.20 *flow diagram of the industrial manufacture of nitric acid*

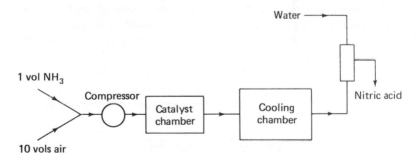

(c) Sulphuric acid

Sulphuric acid is the most widely used chemical of modern manufacturing industry. It is prepared by the contact process in which sulphur dioxide is oxidised by air on a catalyst of vanadium(V) oxide:

$$2SO_2(g) + O_2(g) \rightleftharpoons 2SO_3(g) \qquad \Delta H_r = -594 \text{ kJ mol}^{-1}$$

The application of Le Chatelier's principle to the choice of a moderate running temperature ($500°$ C) was considered in Section 6.4. It was also shown that the yield of sulphur trioxide could be increased with high pressure, but since a pressure fractionally above that of the atmosphere already produces 98% conversion, the extra expense and increased maintenance of high-pressure plant is not justified.

The contact process has the following stages (see Fig. 12.21):

fig 12.21 *flow diagram of the contact process*

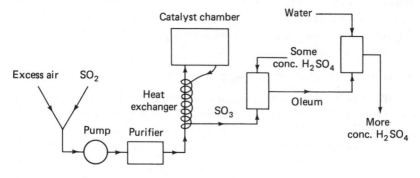

(i) Sulphur dioxide is obtained by burning sulphur in air.

(ii) Sulphur dioxide is mixed with a large excess of air. This excess is necessary because the catalyst is only effective for low sulphur dioxide concentrations.

(iii) The gases are thoroughly purified. Impurities 'poison' the costly catalyst surface.

(iv) The mixture is pumped through a heat exchanger where it is heated to $500°$ C by the hot exhaust gases from the reaction chamber.

(v) In the reaction chamber, sulphur trioxide is formed on the large surface of the catalyst.

(vi) Direct solution of sulphur trioxide in water produces sulphuric acid as a very fine unmanageable mist. In practice, the gas is pumped up a tower against a steady trickle of some existing sulphuric acid to form *oleum*:

$$SO_3(g) + H_2SO_4(l) \longrightarrow H_2S_2O_7(l)$$
$$\text{oleum}$$

(vii) Dilution of oleum then produces more sulphuric acid:

$$H_2S_2O_7(l) + H_2O(l) \longrightarrow 2H_2SO_4(l)$$

(d) Fuel gases

'Gas' is one of the cheapest and most easily handled fuels.

(i) *Natural gas*

The domestic gas supply uses *natural gas*. This is methane, CH_4 (see Section 13.2), trapped under pressure in vast deposits below the bed of the North Sea. Although non-poisonous, a natural gas leak is dangerous because of the explosion risk. For this reason, the gas is impregnated with a foul-smelling chemical so that leaks are quickly detected.

Natural gas can also be obtained from the action of bacteria on raw sewage in the absence of air, although the method has not yet been put to large-scale use.

(ii) *Coal gas*

When coal is heated to $1000°C$ in the absence of air, a mixture of combustible gases is given off, with the typical composition shown in Table 12.9. A light vapour is also obtained which contains a high proportion of ammonia, together with heavy *coal tar* vapour, used widely in the organic chemicals industry (see Section 13.6). Left behind are hard lumps of carbon called *coke*. Coke is used in the reduction of some metal ores, most notably iron oxide in the blast furnace. It is also used to prepare industrial fuel gases.

Table 12.9 *typical volume composition of coal gas*

Gas		Volume composition/%
Hydrogen	H_2	61
Methane	CH_4	18
Carbon monoxide	CO	17
Ethane	C_2H_6	4

(iii) *Fuel gases from coke*

Producer gas

This is obtained when air is blown through hot coke. Oxygen in the air reacts with coke to produce carbon monoxide by an exothermic reaction which rapidly raises the temperature of the coke:

$$2C(s) + O_2(g) \longrightarrow 2CO(g) \qquad \Delta H_r = -220 \text{ kJ mol}^{-1}$$

Nitrogen does not react, but remains in producer gas as a diluent.

Water gas

This is obtained when steam is blown through hot coke. Hydrogen and carbon monoxide form in an endothermic reaction:

$$H_2O(g) + C(s) \longrightarrow H_2(g) + CO(g) \qquad \Delta H_r = +132 \text{ kJ mol}^{-1}$$

Water gas burns with a hotter flame than producer gas, but suffers from the disadvantage that its manufacture requires heat. In practice producer gas and water gas are manufactured simultaneously. The proportions of steam to air are adjusted in order that the heat required for the manufacture of water gas is obtained from the manufacture of producer gas, the temperature of the coke being maintained at approximately 1000°C.

CHAPTER 13

ORGANIC CHEMISTRY

Once upon a time chemicals were neatly divided into two:

Organic chemicals were created by or were part of a living organism. These held *'life force'*.

Inorganic chemicals were derived from rocks and other lifeless objects. These contained no *'life force'*.

But then Wöhler successfully made the organic chemical urea (present in urine) starting with purely inorganic chemicals. Thus it seemed that 'life force' might not exist, that there was no fundamental difference between organic and inorganic chemistry. Nevertheless, the chemicals of life do fall naturally into one class, in which particular elements are linked through similar bonds. Today, in spite of its ill-defined boundaries, organic chemistry is a vast subject in its own right. It is a subject based on carbon, the one element whose atoms can join into long chains, branched chains and rings and at the same time bond with other elements. It is this versatility that makes carbon a good choice for life.

Billions of different organic chemicals are known, and the number grows rapidly each year as more are prepared. Organic chemicals are to be seen walking about everywhere in all of us. They are also in detergents, dyes, plastics, cloth, paper, food, flavourings, glues, smells, plants and rubber. Everyone of these organic chemicals is a compound based upon carbon, whose electron arrangement is

carbon, C   6p 2e 4e

Carbon becomes very stable when it shares four electrons. It has a bonding, also described as a *valency*, of four. Often electrons are shared with other carbon atoms, or with hydrogen atoms to form strong carbon–carbon or carbon–hydrogen bonds. The element is also found bonded to oxygen or to nitrogen atoms, which may themselves be bonded to other atoms. In the organic chemistry met in this chapter

carbon atoms show a bonding, or valency, of 4;
hydrogen atoms show a bonding, or valency, of 1;

oxygen atoms show a bonding, or valency, of 2; and nitrogen atoms show a bonding, or valency, of 3.

13.1 THE PROPERTIES OF ORGANIC CHEMICALS

The backbone of all organic molecules is made up of strong unreactive C–C and C–H covalent bonds. The reactivity of the molecules is governed by the *functional groups*; in other words, by whatever else is added to this basic carbon–hydrogen skeleton. For example, the −O−H functional group affords the following compounds very similar properties.

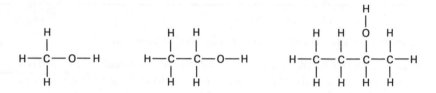

In the same way, the C=C functional group means that the following three compounds behave in a similar manner.

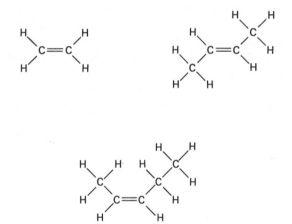

It is therefore logical to classify organic chemicals according to their functional groups. This introduction to the subject examines first the group of organic chemicals with no functional group, and then a small selection of some more common functional groups.

13.2 THE ALKANES

The simplest organic molecules are those containing carbon and hydrogen only. These are the *hydrocarbons*. The simplest of the hydrocarbons are those containing only single covalent C—C bonds. These are *alkanes*.

(a) General

Table 13.1 lists the alkanes containing between one and ten carbon atoms.

Table 13.1   *the alkanes*

Formula	Name	Boiling point/°C
CH_4	methane	− 161
C_2H_6	ethane	−89
C_3H_8	propane	−42
C_4H_{10}	butane	0
C_5H_{12}	pentane	36
C_6H_{14}	hexane	69
C_7H_{16}	heptane	99
C_8H_{18}	octane	126
C_9H_{20}	nonane	151
$C_{10}H_{22}$	decane	174

It can be seen that the

Number of hydrogen atoms =(2 × *number of carbon atoms*) + 2

The general formula for alkanes is therefore C_nH_{2n+2}. The internationally agreed naming system is used here. In this, the first part of the name indicates the number of carbon atoms. Meth- indicates 1, eth- 2, prop- 3, but- 4 carbon atoms, and so on, 'ane' indicates that all C—C bonds are single.

Each carbon atom in an alkane is sharing its four outer electrons with four other atoms, suggesting a tetrahedral arrangement in which all the angles are 109.5° (see Section 4.5). Thus a molecule of pentane should really be drawn as shown in Fig. 13.1. This is often schematically simplified to

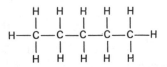

but it should be realised that this is not an accurate representation of shape, and that all bond angles are tetrahedral. Thus both diagrams below represent the same molecule of pentane.

fig 13.1 *a molecule of pentane*

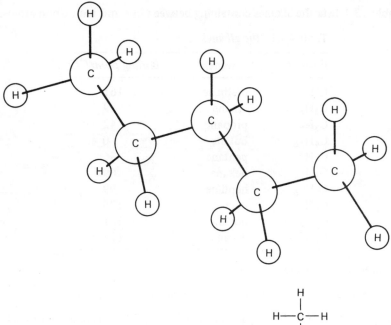

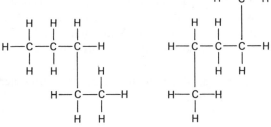

(b) Isomerism

Butane, C_4H_{10}, can have two different structures.

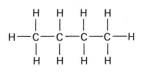

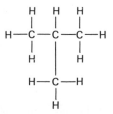

Two different compounds each with the formula C_4H_{10} are known. Compounds with the same formulae but different structures are called *isomers*. Pentane, C_5H_{12}, has three isomers.

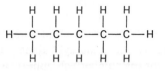

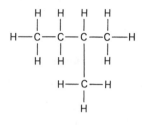

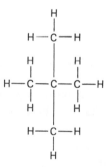

The reader should be able to sketch five isomers for hexane, C_6H_{14}.

(c) The naming of isomers
The longest continuous carbon chain in the structure is selected and numbered.

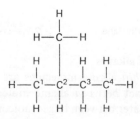

In the above example, the longest continuous chain of four carbon atoms gives the basic name of 'butane'. One

$$\begin{array}{c} H \\ | \\ H-C-H \\ | \end{array}$$

or *methyl group* is bonded to carbon atom number 2 of this chain. The compound is therefore named '2-methyl-butane'. The numbering of the chain is such that the lowest number will appear in the name. The isomer of hexane, C_6H_{14}, shown below also has four carbon atoms in its longest continuous chain.

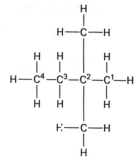

Two methyl groups are attached to carbon atom number 2 of this chain. '*Di*methyl' indicates that there are two such groups, and again the point of attachment of each group is indicated by numbers. Hence the name of this isomer is '2,2-dimethyl-butane'. Further examples are given below.

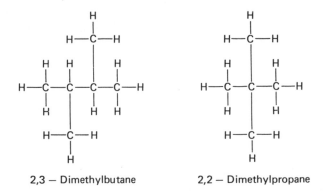

2,3 – Dimethylbutane   2,2 – Dimethylpropane

(d) Physical properties of alkanes

Although the bonds *within* alkane molecules are very strong, only very weak secondary bonds act *between* alkane molecules (see Section 3.8). Secondary bonding is greater between larger molecules. Thus

(i) CH_4 to C_4H_{10} inclusive are gases,

(ii) C_5H_{12} to $C_{16}H_{34}$ inclusive are oily liquids, becoming more viscous as the chain length increases, and

(iii) $C_{17}H_{36}$ onwards are waxy solids.

Boiling point, melting point and heat of fusion show a steady increase with increased carbon chain length, because each of these values is dependent upon how strongly the molecules are held together.

The alkanes are colourless and, when pure, have no smell. They are non-polar (their molecules have no slightly positive nor slightly negative ends) and they are insoluble in polar solvents such as water.

(e) Chemical reactions

The very strong C–C and C–H covalent bonds afford alkanes their main feature of non-reactivity. All that alkanes do is (i) to burn in air and oxygen and (ii) to react with halogens in ultra-violet light.

(i) *Burning*

The carbon and hydrogen in most organic compounds is more stable as carbon dioxide and water vapour. Thus alkanes burn in air with a clear flame in a strongly exothermic reaction. For this reason they make good fuels:

$$CH_4(g) + 2O_2(g) \longrightarrow CO_2(g) + 2H_2O(l) \qquad \Delta H_c = -890 \text{ kJ mol}^{-1}$$

$$C_7H_{16}(g) + 11O_2(g) \longrightarrow 7CO_2(g) + 8H_2O(l) \qquad \Delta H_c = -4853 \text{ kJ mol}^{-1}$$

Liquid alkanes must vaporise before they will burn effectively. If the alkane chain is long, or if there is an insufficient supply of air, then poisonous carbon monoxide or even free carbon (soot) is produced in place of carbon dioxide. In this case, the flame is more yellow and luminous:

$$CH_4(g) + 1\tfrac{1}{2}O_2(g) \longrightarrow CO(g) + 2H_2O(l)$$
limited
supply of
air

It is the incomplete combustion of the alkanes in petrol and diesel which is responsible for over 900 000 tonnes* of carbon monoxide being pumped into London's atmosphere each year. Similarly, it is dangerous to cover the air intake to a gas boiler, since limiting the supply of air produces poisonous carbon monoxide in the combustion of natural gas (methane).

*Extrapolated value for 1980s following conversation with Dr Apling of Warren Springs Laboratories, Stevenage, Herts.

(ii) *Reaction with halogens*

If a few drops of red–brown liquid bromine are added to a liquid alkane, such as hexane, in a test tube, the brown colour of the bromine persists. If two such mixtures are prepared and one is left in a dark cupboard for half an hour whilst the other is kept in daylight, the latter gradually loses its colour, whilst the hidden solution remains brown. Bromine reacts with alkanes only in the presence of ultra-violet light. The pulses of light set bromine molecules vibrating until they vibrate apart into separate atoms,

fig 13.2 *ultra-violet light causes bromine molecules to vibrate apart to form reactive bromine atoms*

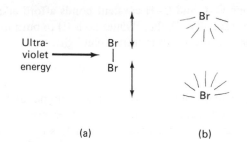

(a) (b)

as shown in Fig. 13.2. The free atoms are sufficiently reactive to attack the alkane. The overall reaction is represented by the equation:

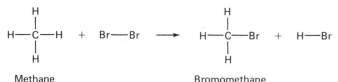

Methane   Bromomethane

This is a *substitution* reaction. In this case, bromine has been substituted for hydrogen. The reaction does not stop at forming bromo-methane, but further substitution occurs:

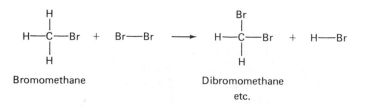

Bromomethane   Dibromomethane
                                                     etc.

Chlorine reacts in a similar manner.

(f) Homologous series

All alkanes behave in a similar way. A group of chemicals with similar chemical properties, a gradation of physical properties and the same general formula are said to form a *homologous series*.

(g) The alkanes in everyday life
Many members of this homologous series are constituents of fuels (see Table 13.2). Longer-chain alkanes are used as lubricating oils or as waxy

Table 13.2 *typical alkane fuels*

Fuel	Number of carbon atoms in alkane molecule
Natural gas	1
Calor gas	4
Petrol	5-10
Pink or blue paraffin	10-16
Diesel	approx. 17
Candle wax	approx. 28

waterproofing agents for paper drinking cartons and the like. Today a large proportion of the alkanes obtained from crude oil is converted to more reactive alkenes (see following section) from which almost every industrial organic product can be manufactured.

13.3 THE ALKENES

Alkenes are hydrocarbons containing doubly bonded carbon atoms.

(a) General
The simplest alkene, ethene, contains two carbon atoms; the next simplest, propene, contains three; and so on.

Ethene

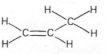

Propene

Table 13.3 lists some of the simpler alkenes. The general formula is C_nH_{2n}. The doubly bonded carbon atoms can be seen to bond with three other atoms, giving rise to a planar structure with an angle of 120° between the bonds (see Section 3.5).

Table 13.3 *the alkenes*

Formula	Name
C_2H_4	ethene
C_3H_6	propene
C_4H_8	butene
C_5H_{10}	pentene
C_6H_{12}	hexene

(b) Naming and isomerism

There are three isomers of butene, C_4H_8. A number placed before '-ene' in the name specifies the carbon atom upon which the double bond starts.

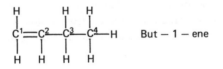

But − 1 − ene

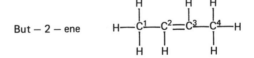

But − 2 − ene

2-Methylpropene

(c) Physical properties

Like alkanes, melting point, boiling point, heat of fusion and viscosity increase with increasing carbon chain length.

(d) Chemical reactions

(i) With the halogens

If a solution of bromine in water or in tetrachloromethane is added to an alkene, the brown colour disappears instantly, even in the absence of daylight:

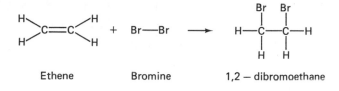

| Ethene | Bromine | 1,2 – dibromoethane |

Similar reactions occur with chlorine or iodine.

(ii) With hydrogen

A mixture of an alkene and hydrogen reacts when passed over a heated nickel catalyst. The product is an alkane:

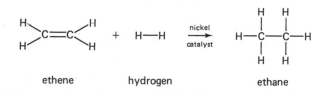

| ethene | hydrogen | ethane |

The reactivity of alkenes is explained by the ease with which the double bond breaks open.

In the above reactions, halogens or hydrogen add on as the bond breaks open. These are called *addition reactions*. The alkene molecule is said to be *unsaturated*, meaning that it can undergo addition, in the process becoming *saturated*. Any molecule containing a double bond is unsaturated.

(iii) *Polymerisation*

Many alkene molecules can be persuaded to add to one another to form long chains. The product of linking many molecules is a *polymer* (Fig. 13.3). This is *addition polymerisation*. This polymerisation, which was

fig **13.3** *addition polymerisation*

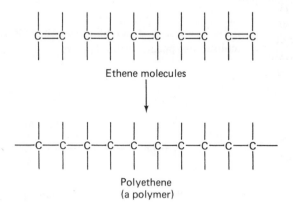

discovered quite by accident, occurs when ethene is compressed in the presence of a catalyst.

The polymerisation of similar unsaturated molecules produces a number of well known products (see Section 14.3(a)).

(e) Homologous series

Like the alkanes, the alkenes show similar chemical properties, a gradation of physical properties and the same general formula. The alkenes therefore make up another homologous series.

(f) Manufacture of alkenes

Alkenes are less stable than alkanes. Energy must be given to alkanes in their conversion to alkenes.

Industrially the process of *cracking* creates an assortment of alkenes from long-chain alkanes. Energy is given to an alkane vapour by passing it over a large, hot surface area (heated porous brick provides such a surface). The experiment in Fig. 13.4 illustrates cracking. It is important to remove the tube from the beaker of water as soon as heating is stopped in order to prevent cold water being sucked back into the hot glass test tube.

The *results* of the experiment can thus be written as:

(i) A *gas* is collected. If the tube of this gas is lit, it burns steadily. This suggests that shorter-chain organic molecules have been produced (smaller molecules are likely to be gaseous).

fig 13.4 *the cracking of a long-chain alkane*

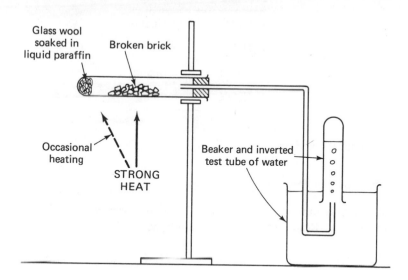

(ii) If aqueous bromine is shaken with the gas, the colour disappears instantly, suggesting alkenes to be present.

(g) Alkenes in everyday life

The use of alkenes in manufacturing addition polymers has been mentioned. Because of their greater reactivity, the alkenes are manufactured from alkanes as an essential step in the production of most commercially important organic chemicals—solvents, fibres, paints, drugs, flavourings.

13.4 THE ALCOHOLS

Alcohol is quite a lot more than the sociable constituent of spirits, beers and wines. Any hydrocarbon in which one H atom has been substituted by an —O—H group is an alcohol, and alcohols form another *homologous series.*

(a) General

Table 13.4 lists some of the simpler alcohols. The O—H group of alcohols is polar, the oxygen being slightly negative and hydrogen slightly positive. Thus alcohol molecules are attracted to one another through hydrogen bonds, the strongest type of secondary bond (see Section 3.8(b)). Thus even methanol is a liquid at ordinary temperatures.

Table 13.4 *the alcohols*

Formula	Name
CH_3OH	methanol
C_2H_5OH	ethanol
C_3H_7OH	propanol
C_4H_9OH	butanol

(b) Chemical reactions

(i) *Burning*

The alcohols burn in air with clear flames to form carbon dioxide and water vapour. Reactions are highly exothermic

$$C_2H_5OH(l) + 3O_2(g) \longrightarrow 2CO_2(g) + 3H_2O(l) \quad \Delta H_c = -1367 \text{ kJ mol}^{-1}$$

and for this reason the alcohols make excellent fuels.

(ii) *With phosphorus pentachloride*

When powdered phosphorus pentachloride, PCl_5, is dropped into a sample of any liquid containing the $-O-H$ group, a cloud of hydrogen chloride gas, HCl, immediately forms:

$$C_2H_5OH(l) + PCl_5(s) \longrightarrow \underset{\text{chloroethane}}{C_2H_5Cl(alc)} + \underset{\substack{\text{phosphorus} \\ \text{trichloride} \\ \text{oxide}}}{POCl_3(alc)} + HCl(g)$$

The presence of hydrogen chloride is easily verified by the dense white smoke that forms when a drop of aqueous ammonia is held near (see Section 12.5(c)). This is used as a test for the presence of an $-O-H$ group.

(iii) *With oxidising agents*

The alcohol of alcoholic drinks is ethanol, C_2H_5OH. A bottle of wine which has been left uncorked for some days will often taste of vinegar, vinegar being a dilute solution of ethanoic acid. The ethanol in the wine has been oxidised to ethanoic acid:

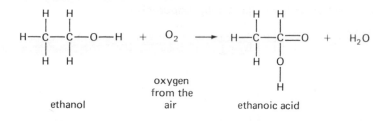

oxygen
from the
ethanol   air   ethanoic acid

In fact, tiny living organisms present in air catalyse this reaction. Whisky, gin and other spirits do not oxidise to vinegar when left open to the air because the greater concentration of alcohol in these drinks (about 40% in whisky) kills these micro-organisms before they can induce reaction.

The oxidation of methanol by air can be catalysed simply by a warm platinum wire. If held in the vapour over a beaker of methanol, it begins to glow red hot as the exothermic oxidation proceeds (Fig. 13.5).

fig 13.5 *oxidation of methanol catalysed by warm platinum wire*

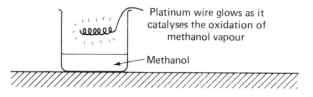

Platinum wire glows as it catalyses the oxidation of methanol vapour

Methanol

If an orange solution of potassium dichromate(VI) in dilute sulphuric acid is warmed gently with ethanol, the solution begins to darken and eventually turns olive-green. In order to prevent the ethanol from evaporating away during heating, escaping vapour is condensed and returned to the flask as shown in Fig. 13.6. This arrangement is called *refluxing*.

Acidified potassium dichromate(VI) solution is an oxidising agent, turning from orange to green during oxidation. After half an hour's refluxing the contents of the flask smell of vinegar or ethanoic acid. Again, ethanol has been oxidised to ethanoic acid.

(c) The production of alcohols

(i) *Methanol*

(1) From wood

Methanol is one of many products obtained when wood is heated in the absence of air (Fig. 13.7). The condensed liquid contains some methanol. If it is warmed with dilute acidified potassium dichromate(VI) solution, the colour changes from orange to green.

fig **13.6** *refluxing to prevent loss by evaporation*

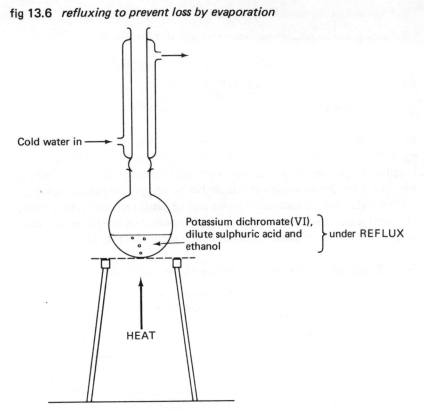

Cold water in ⟶

Potassium dichromate(VI),
dilute sulphuric acid and }under REFLUX
ethanol

HEAT

fig '**13.7** *methanol from the distillation of wood*

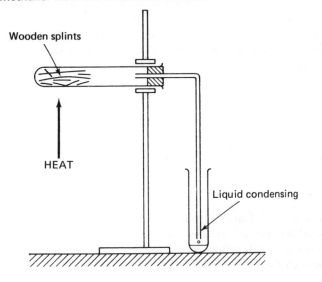

Wooden splints

HEAT

Liquid condensing

(2) From carbon monoxide and hydrogen

Methanol is produced commercially by the following reaction:

$$CO(g) + 2H_2(g) \longrightarrow CH_3OH(l)$$

Both a catalyst and high pressure are necessary.

(ii) *Ethanol*

(1) From sugar or starch

Ethanol is obtained by the biological action of yeast on aqueous solutions of sugar or starch. Yeast contains biological catalysts called *enzymes* which work only at very precise temperatures. The reaction can be demonstrated by adding a few granules of dried baker's yeast to a flask of sugar solution (Fig. 13.8). The mixture should be kept at about 25°C (perhaps above a

fig 13.8 *the preparation of ethanol by fermentation*

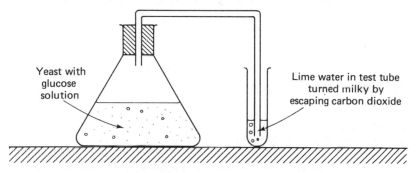

Yeast with glucose solution

Lime water in test tube turned milky by escaping carbon dioxide

radiator or in an airing cupboard) for some days. Carbon dioxide steadily bubbles off as ethanol forms in the flask:

$$\underset{\text{glucose sugar}}{C_6H_{12}O_6(aq)} \xrightarrow[25°C]{\text{in presence of yeast}} \underset{\text{ethanol}}{2C_2H_5OH(aq)} + 2CO_2(g)$$

This is the process occurring in wine making, sugar being present in the juice of the grape, orange or whatever other wine is being prepared. The process is called *fermentation*.

(2) From ethene and water

Most alcohol for non-drinking purposes is manufactured from ethene and water vapour, which react under pressure in the presence of a catalyst of phosphoric(V) acid:

$$C_2H_4(g) + H_2O(g) \longrightarrow C_2H_5OH(l)$$

(d) Alcohols in everyday life

It has already been mentioned that alcohols provide more in our everyday lives than the odd drink.

(i) As fuels

Higher grades of petrol contain some ethanol. This alcohol is also injected into aviation fuel to improve the jet boost on take-off. *Methylated spirits* is mainly ethanol, contaminated with poisonous methanol, pink dye and foul-tasting pyridine to make it undrinkable.

(ii) In medicine

Neat ethanol kills germs. As *surgical spirit* it is used to sterilise wounds. Methanol and ethanol are used in the manufacture of many drugs.

(iii) As solvents

Industrial methylated spirits (ethanol contaminated with 5% methanol) is used as a solvent for some varnishes and glues.

(iv) Other uses

The double alcohol ethan-1,2-diol (glycol) is used as an antifreeze in car radiators, and in the manufacture of Terylene (see Section 14.2(b)). Alcohols are used to manufacture alkanoic acids which themselves find a wide industrial application (see next section).

13.5 ORGANIC ACIDS

Organic acids are the oxidation products of alcohols.

Just as the homologous series of alcohols contained the $-O-H$ group, so the homologous series of *alkanoic acids* contains the

or $-CO_2H$ group.

(a) General

Table 13.5 lists the simpler alkanoic acids. As with the alcohols, hydrogen bonding holds neighbouring molecules together. Thus even methanoic acid is a liquid. The acids have very strong tastes and smells. It is butyric acid

Table 13.5 *the alkanoic acids*

Formula	Name
HCO_2H	methanoic acid
CH_3CO_2H	ethanoic acid
$C_2H_5CO_2H$	propanoic acid
$C_3H_7CO_2H$	butanoic acid

which gives the strong taste to some cheeses, and dilute ethanoic acid is vinegar.

(b) Chemical reactions

(i) *With water*

The alkanoic acids are *weak acids*, dissociating only slightly in water to form hydrogen ions and *alkanoate* ions:

$$CH_3CO_2H(aq) \rightleftharpoons CH_3CO_2^-(aq) + H^+(aq)$$

ethanoic   ethanoate   hydrogen
acid   ion   ion

Nevertheless, they do show the following typical acid properties.
 (i) They turn universal indicator solution yellow or red.
(ii) They react with metals high in the electrochemical series to form the metal salt and hydrogen gas:

$$Zn(s) + 2CH_3CO_2H(aq) \longrightarrow Zn(CH_3CO_2)_2(aq) + H_2(g)$$

ethanoic   zinc ethanoate
acid   (contains Zn^{2+}
                and $CH_3CO_2^-$ ions)

(iii) They react with carbonates to give carbon dioxide gas:

$$CaCO_3(s) + 2CH_3CO_2H(aq) \longrightarrow Ca(CH_3CO_2)_2(aq) + H_2O(l) + CO_2(g)$$

ethanoic   calcium ethanoate
acid   (contains Ca^{2+} and
                $CH_3CO_2^-$ ions)

(iv) They react with bases to form a salt and water:

$$CH_3CO_2H(aq) + NaOH(aq) \longrightarrow Na(CH_3CO_2)(aq) + H_2O(l)$$

acid   base   salt   water

sodium ethanoate
(contains Na^+ and
$CH_3CO_2{}^-$ ions)

(ii) *With phosphorus pentachloride*

Alkanoic acids, like alcohols, contain the $-O-H$ group. They, too, react spontaneously with phosphorus pentachloride to form a cloud of hydrogen chloride gas (see Section 13.4 (b)).

(iii) *With alcohols*

This reaction can be examined experimentally by mixing a little of any alkanoic acid with a slightly greater volume of any alcohol and two drops of concentrated sulphuric acid in a test tube. The mixture is then warmed by standing in a beaker of hot water for a few minutes.

If the sulphuric acid is now diluted by pouring the mixture into an equal volume of cold water in a second beaker, a pleasant fruity smell is generally obtained. Table 13.6 shows the results of some alkanoic acid-alcohol reactions.

Table 13.6   *results of some alkanoic acid–alcohol reactions*

Alkanoic acid	Alcohol	Smell
ethanoic acid	butanol	boiled sweet
methanoic acid	ethanol	rum
ethanoic acid	ethanol	glue

The smell shows the presence of *esters*, formed by the reversible reaction between alkanoic acids and alcohols:

alkanoic acid + alcohol $\rightleftharpoons$ ester + water

In the two examples given in Fig. 13.9, the parts of the molecules which break away to form water are ringed. These reactions are catalysed by $H^+(aq)$ ions. The addition of concentrated sulphuric acid to the reaction mixture, in addition to supplying the $H^+(aq)$ catalyst, removes water (the

fig 13.9 *two alkanoic acid–alcohol reactions*

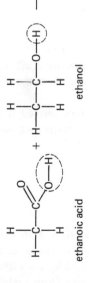
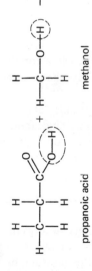
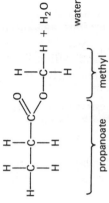
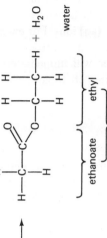

acid is a dehydrating agent) according to Le Chatelier's principle (see Section 6.4).

(iv) *The hydrolysis of esters*

Since the above reactions are reversible, heating an ester–water mixture will produce alkanoic acid and alcohol. The reaction of a substance with water is known as *hydrolysis*. The hydrolysis reaction of an ester is catalysed by H^+(aq) ions, and also by OH^-(aq) ions. For example, boiling the ester ethylethanoate with aqueous sodium hydroxide produces ethanol and ethanoic acid, although the latter will immediately react with the sodium hydroxide catalyst to form sodium ethanoate.

(c) Acids and esters in everyday life

(i) *Acids*

Butanoic and pentanoic acids give part of the strong flavour to some margarines, butter and cheese. Ethanoic acid solution is vinegar. This acid is also used in the production of rayon from cellulose.

Methanoic acid is in the sting of ants.

The salts of long-chain alkanoic acids are soaps (see Section 14.2).

(ii) *Esters*

Esters are widely used as solvents for glues and varnishes. Naturally occurring esters are responsible for the flavours of many fruits and the smells of many flowers. Synthetic esters are used as flavouring additives and perfumes.

Animal and vegetable fats and oils are esters derived from long-chain alkanoic acids (containing about 18 carbon atoms per molecule) combined with the triple alcohol, propan-1,2,3-triol.

13.6 THE SOURCE OF ORGANIC CHEMICALS

Most of the millions of organic compounds that Man has synthesised come from the basic organic compounds found in crude oil or coal. Crude oil results from the pressure of sediment which has settled upon the remains of billions of microscopic sea organisms from prehistory. It contains many simple hydrocarbon chains. Coal is formed by the pressure of sediment which has settled upon the remains of prehistoric forests. It consists of carbon together with more complex hydrocarbon rings and ammonia. The two schemes in Figs 13.10 and 13.11 give an indication of

fig 13.10 *products from coal*

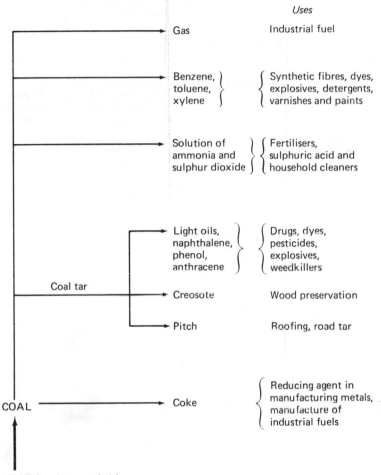

Uses

Gas — Industrial fuel

Benzene, toluene, xylene — Synthetic fibres, dyes, explosives, detergents, varnishes and paints

Solution of ammonia and sulphur dioxide — Fertilisers, sulphuric acid and household cleaners

Light oils, naphthalene, phenol, anthracene — Drugs, dyes, pesticides, explosives, weedkillers

Coal tar

Creosote — Wood preservation

Pitch — Roofing, road tar

COAL — Coke — Reducing agent in manufacturing metals, manufacture of industrial fuels

HEAT (in absence of air)

the value of these two commodities. It can be seen that coal is not just for burning on an open fire, and that crude oil provides more than just a source of petrol.

fig **13.11** *products from crude oil*

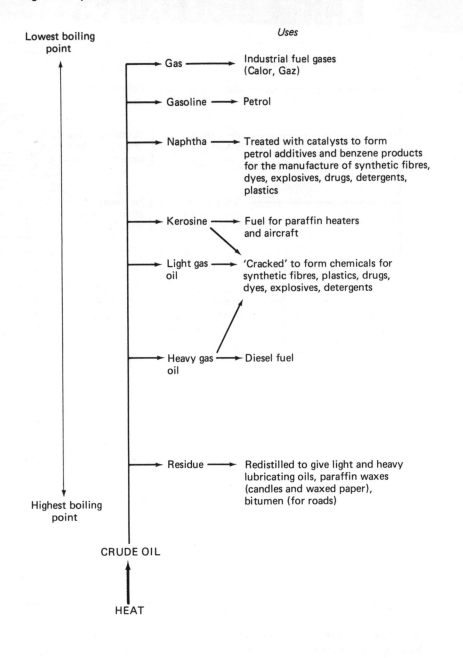

CHAPTER 14

LARGE MOLECULES

Many naturally occurring molecules are large. One molecule of an animal fat contains 173 atoms; a molecule of the major component of linseed oil contains 161 atoms. In spite of their size, each molecule has a precise structure and a definite position for each one of its atoms. Other naturally occurring molecules are exceptionally large. Starch molecules contain between 1500 and 150 000 atoms; rubber molecules between 13 000 and 65 000 atoms. In these examples, it is necessary to give a range of values because different molecules of the same substance have different lengths. Such molecules are built up by linking hundreds of small molecules into one long chain. These are *polymers* in which the chain length is quite variable. This chapter first considers some large molecules with a precise formula. These are the fats and oils, and the detergents derived from them. Secondly, polymers are examined.

14.1 FATS AND OILS

An ester (see Section 13.5(b)) is derived from the reversible reaction between an alkanoic acid and an alcohol:

$$\text{alkanoic acid} + \text{alcohol} \rightleftharpoons \text{ester} + \text{water}$$

A fat is a naturally occurring ester derived from a long-chain alkanoic acid and a triple alcohol. In the same way that a simple alcohol molecule reacts with one alkanoic acid, so a triple alcohol reacts with three. The triple alcohol, or tri-ol, is propan-1,2,3-triol (often called glycerol or glycerine) and the long-chain acid could be octadecanoic acid (or stearic acid), $C_{17}H_{35}CO_2H$. Their reaction to form an ester is shown diagrammatically in Fig. 14.1. Atoms which break off to form water are ringed.

All the various animal and vegetable fats and oils are esters with this basic structure. They differ only in the alkanoic acid part of the molecule. For example, the esters in palm oil are derived principally from

fig 14.1 *reaction between a long-chain acid and a triol*

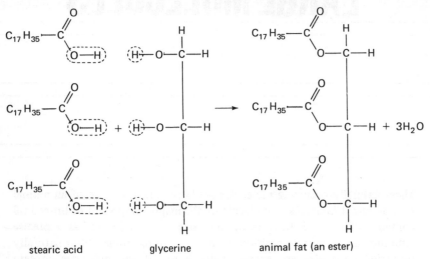

| stearic acid | glycerine | animal fat (an ester) |

hexadecanoic acid (or palmitic acid) $C_{15}H_{32}CO_2H$. The long chain of some fat esters contain several $C=C$ double bonds. Such molecules are said to be *unsaturated* (see Section 13.3(d)), and these are the *polyunsaturated fats* mentioned on the side of 'health' margarines. Such fats are claimed to reduce the chance of heart disease. Vegetable and fish oils and fats are generally high in polyunsaturated fats, whilst animal fats are saturated.

14.2 DETERGENTS

There are two types of detergents: *soapy* and *non-soapy*. In everyday language, soapy detergents are 'soaps' and non-soapy detergents are just 'detergents'.

(a) Soaps

A soap is the soluble salt, usually the lithium, sodium or potassium salt, of a long-chain alkanoic acid. For simplicity, the long hydrocarbon chain of such an acid is represented in the following diagrams by a zig-zag line. Thus stearic acid, $C_{17}H_{35}CO_2H$, is represented as

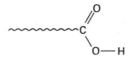

(i) *Manufacture of soap*

Section 13.5(b) showed that alkanoic acid and alcohol are obtained when

an ester is hydrolysed. If sodium hydroxide is added to catalyse the reaction, then it will react with the alkanoic acid produced, forming its sodium salt as the final product:

$$\text{ester + water} \xrightarrow[\text{NaOH}]{\text{catalysed by}} \text{acid + alcohol}$$

reacts
with
NaOH

sodium salt of
acid

Fats are esters of long-chain alkanoic acids (see above). Thus boiling a fat with sodium hydroxide, or any other alkali, produces the salts of long-chain acids, and these are soaps. The process is called *saponification* (Fig. 14.2).

fig 14.2 *the hydrolysis of fat by alkali*

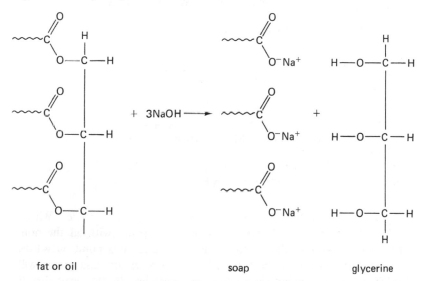

fat or oil   soap   glycerine

A fact not mentioned in expensive soap advertisements is that many soaps are the product of boiling the esters in molten pig's fat with caustic soda drain cleaner.

The commercial manufacture of soap can be illustrated in the laboratory by carefully boiling and stirring a mixture of moderately concentrated sodium hydroxide solution with half its volume of castor oil for about ten minutes. If the mixture is then left to cool a solid crust of soap collects on the surface. This can be scraped off and washed in running water to remove excess alkali. An increased yield of soap is obtained by *salting out*, which involves stirring common salt into the mixture before cooling. This helps to push more soap out of solution.

If potassium hydroxide is used, liquid soaps are obtained.

(ii) *Action of soap*

Soap enables dirt and grease to be washed away in water, when water alone has no effect. Water molecules are *polar*; they have one slightly negative and one slightly positive end. For a substance to be attracted into, and therefore dissolved into, water, it too must be polar. It can then be pulled into the water by electrical attraction. Dirt and grease, being mainly organic hydrocarbons, are non-polar and therefore do not dissolve in water. They only dissolve in non-polar solvents such as hydrocarbon alkanes.

Soap has a non-polar hydrocarbon end (this will dissolve grease) and a polar charged end (this will dissolve in water). Each soap molecule will be represented as a tadpole with a polar head, P (Fig. 14.3).

fig **14.3** *the structure of soap*

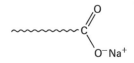

Non-polar (oily) end   Polar end
(dissolves with oils)   (dissolves in water)

Agitating soap with water produces millions of tiny *micelles*. Within each micelle the soap molecules are grouped together with all the non-polar ends together in the centre whilst the polar ends point outwards, attracted towards the water (Fig. 14.4). Non-polar dirt and grease will dissolve in the non-polar centre of these micelles. In this way dirt is indirectly dissolved in water.

(b) Detergents

Detergents act in the same way as soaps. Structurally they are similar, with both polar and non-polar ends, but with the difference that the polar part of the detergent molecule is commonly a *sulphate* group or a *sulphonate* group (Fig. 14.5). Such detergents are manufactured by treating the long-

fig 14.4 *soap molecules in water*

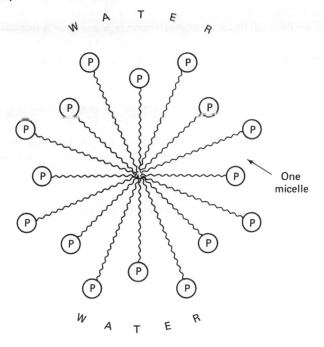

fig 14.5 *the structure of detergent*

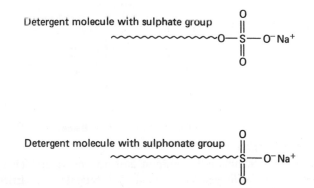

chain hydrocarbons obtained from crude oil with concentrated sulphuric acid.

Washing powders actually contain only about 20% detergent. The remaining 80% typically comprises phosphate to soften the water together with a mild bleaching agent and sodium sulphate as a powdering agent.

14.3 POLYMERS

When many small molecules bond together, a *polymer* is formed. Polymers are often classified according to the means by which they were formed. For this reason, this section begins by considering synthetic polymers.

(a) Addition polymers
The addition polymerisation of ethene to polyethene (polythene) has already been mentioned in Section 13.3(d). Other alkene molecules similarly add to one another, their C=C double bonds breaking open in the presence of suitable catalysts. The general reaction is summarised in Fig. 14.6, as in the previous chapter. A more economic convention for

fig 14.6 *addition polymerisation*

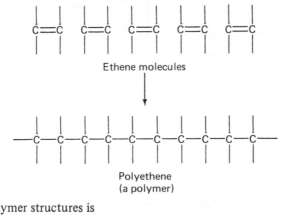

Ethene molecules

Polyethene
(a polymer)

writing polymer structures is

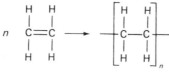

ethene   polyethene

where n is a large number.

Table 14.1 summarises common addition polymers. These polymerisations, which are in general initiated by a catalyst, are carried out industrially under moderate pressure.

(b) Condensation polymerisation
In a *condensation reaction*, molecules link together following removal of water from between them. *Condensation polymerisation* results from the repeated elimination of water in this way. The example in Fig. 14.7 shows condensation between a double alkanoic acid and a double alcohol. The

Table 14.1 *some common addition polymers*

Starting material		Polymer	
Structure	Name	Structure	Name
$\begin{array}{cc} H & H \\ \mid & \mid \\ C & = C \\ \mid & \mid \\ H & H \end{array}$	ethene	$\left[\begin{array}{cc} H & H \\ \mid & \mid \\ -C & -C- \\ \mid & \mid \\ H & H \end{array}\right]_n$	polythene
$\begin{array}{cc} H & Cl \\ \mid & \mid \\ C & = C \\ \mid & \mid \\ H & H \end{array}$	vinyl-chloride	$\left[\begin{array}{cc} H & Cl \\ \mid & \mid \\ -C & -C- \\ \mid & \mid \\ H & H \end{array}\right]_n$	PVC
$\begin{array}{cc} H & CH_3 \\ \mid & \mid \\ C & = C \\ \mid & \mid \\ H & CO_2CH_3 \end{array}$	methyl-methacrylate	$\left[\begin{array}{cc} H & CH_3 \\ \mid & \mid \\ -C & -C- \\ \mid & \mid \\ H & CO_2CH_3 \end{array}\right]_n$	Perspex
$\begin{array}{cc} H & C_6H_5 \\ \mid & \mid \\ C & = C \\ \mid & \mid \\ H & H \end{array}$	styrene	$\left[\begin{array}{cc} H & C_6H_5 \\ \mid & \mid \\ -C & -C- \\ \mid & \mid \\ H & H \end{array}\right]_n$	polystyrene
$\begin{array}{cc} F & F \\ \mid & \mid \\ C & = C \\ \mid & \mid \\ F & F \end{array}$	tetrafluoroethene	$\left[\begin{array}{cc} F & F \\ \mid & \mid \\ -C & -C- \\ \mid & \mid \\ F & F \end{array}\right]_n$	PTFE or Teflon

double acid is terephthalic acid, and the double alcohol is ethan-1,2-diol.

Since acid and alcohol react to form ester, the resulting condensation polymer is a *polyester* called, in this example, *Terylene*.

Nylon, so-called because it was developed in *N*ew *Y*ork and *Lon*don, is another condensation polymer. In this case the condensation is between the double acid hexan-1,6-dioic acid (adipic acid) and the double *amine*, 1,6-diaminohexane (Fig. 14.8). This is a *polyamide*.

Nylon can be prepared in the laboratory by pouring an aqueous solution of the amine on top of a dense solution of the chloride of adipic acid in tetrachloromethane. A skin of nylon forms between the two layers, and this can be fished out to produce a continuous thread of nylon (Fig. 14.9).

248

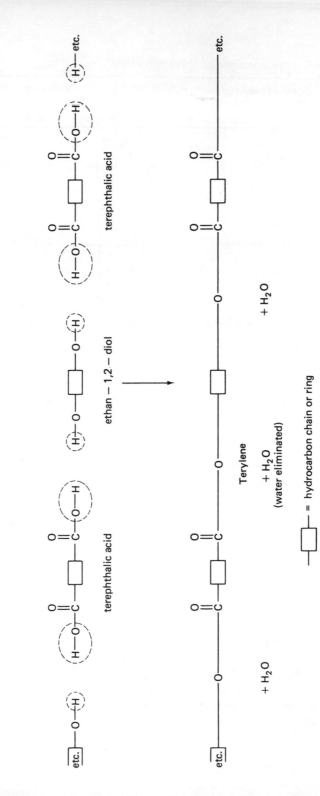

fig 14.7 the formation of a polyester by a condensation reaction

terephthalic acid

ethan − 1,2 − diol

terephthalic acid

Terylene

+ H_2O
(water eliminated)

+ H_2O

+ H_2O

etc.

= hydrocarbon chain or ring

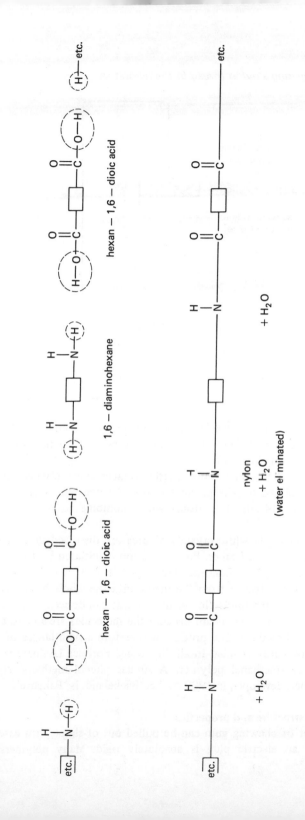

fig 14.8 *the formation of nylon by condensation*

250

fig 14.9 *preparing a nylon thread in the laboratory*

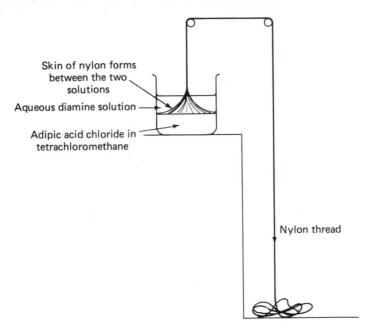

Skin of nylon forms between the two solutions

Aqueous diamine solution

Adipic acid chloride in tetrachloromethane

Nylon thread

The long-chain molecules of nylon and Terylene contain around 100 units. They twist and wrap around one another to produce long fibres suitable for weaving into cloth.

Very much harder and more brittle condensation polymers are also known, exemplified by the product obtained when methanal is reacted with urea (Fig. 14.10). The atoms which combine to form water are ringed.

In the laboratory, white crystals of urea are dissolved in 50 cm³ of aqueous methanal in a beaker. Polymerisation is initiated by the addition of two or three drops of concentrated sulphuric acid. A white solid mass begins to spread outwards from the drops until the whole beaker is filled with a smooth crisp white solid. In this case, it can be seen that the polymer does not form as one chain because the molecules are able to branch off in various directions. The product is therefore a rigid lattice of atoms which are covalently bonded in all directions. Formica laminate is based upon this urea–methanal polymer. A similar phenol–methanal condensation polymer, developed in 1909 by Leo Baekeland, is 'Bakelite'.

(c) Polymer structure and properties

The polymer of chewing gum can be pulled out of shape with ease. The polymer of an electric plug is absolutely rigid. Many polymers have

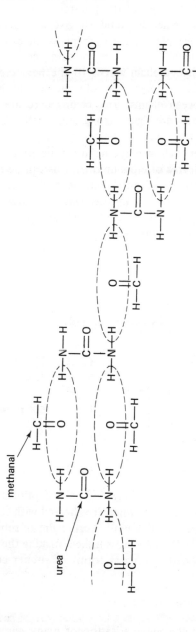

fig 14.10 *the formation of urea–methanal polymer*

properties somewhere in between these two extremes. A polythene shopping bag, for example, must not be too stretchy or the shopper would end up dragging the potatoes home along the pavement, but it must not be too rigid and brittle or the weight of the potatoes would crack the bag open.

Simple long-chain molecules tend to give an easily stretched, easily deformed (i.e. *plastic*) polymer. Polythene is an example of this type.

A polymer in which the covalent bonding continues not just along chains but also from one chain across to the next creates a giant structure. To stretch such a polymer would require the breaking of strong covalent bonds between chains. The chains here are said to be *cross-linked*. Such polymers are rigid, strong and brittle. Urea–methanal provides an example of this type.

A plastic polymer such as polyethene can be strengthened by allowing a small controlled degree of cross-linking. This can be readily achieved by adding a low percentage of but-1,3-diene, $CH_2=CH-CH=CH_2$, to the ethene before polymerisation. In this way the strength, plasticity and brittleness of a polymer can be controlled, and polymers can be synthesised with precisely the required properties.

(d) Natural polymers

(i) Rubber

Rubber is a long-chain hydrocarbon polymer containing C=C double bonds:

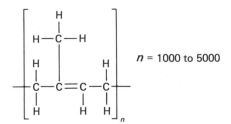

$n = 1000$ to 5000

Natural rubber is not very strong and is exceptionally 'squashy'. It is strengthened by *vulcanisation*, when it is heated with sulphur. The double bonds then break open and link with short chains of sulphur atoms. In this way, rubber molecules become cross-linked, making the rubber harder and less plastic. The hardest, strongest forms of rubber contain up to 35% sulphur.

(ii) Proteins

Protein results from the polymerisation of many *amino acid* molecules.

H—N—C—C—O—H   An amino acid molecule

(with H, H, O above the chain and R below the C)

There are twenty-two amino acids found in life, each differing from the others in the nature of R. For example,

(i) if R is H, the amino acid is *glycine*;

(ii) if R is CH_3, the amino acid is *alanine*.

Polymerisation occurs through condensation (Fig. 14.11). The atoms which combine to eliminate water are ringed.

A particular protein will contain a definite sequence of the different amino acid units peculiar to itself. The protein chains are highly polar, and strong hydrogen bonds (see Section 3.8(b)) hold one chain to another. Some amino acid units contain sulphur atoms in the R group, and these can form S—S covalent cross-links between one protein and the next. In this way strong solid proteins are created, as in fingernail, hoof, skin, wool and silk. When hair is 'permed' a reducing agent is rubbed in which breaks S—S bonds within each hair. After shaping, the addition of an oxidising agent reforms the bonds in new positions. The long molecules of egg white are compact because hydrogen bonding holds each molecule to itself in a tight coil. The heat of cooking is sufficient to vibrate the coils open, enabling them to bond with other molecules in a randomly attracted mass. In this way the egg is cooked solid.

(iii) *Glucose polymers*

Glucose is a sweet-tasting simple sugar with the formula $C_6H_{12}O_6$. The formula shows that, like water, it contains twice as many atoms of hydrogen as oxygen. It is therefore described as a hydrate of carbon, or a *carbohydrate*. Each molecule contains five —O—H groups, and these are involved in a *condensation polymerisation* to produce the carbohydrates cellulose or starch (Fig. 14.12). The atoms which combine to form water are ringed.

These polymerisations can be accomplished by plants; Man has been unable to mimic the reaction.

Cellulose and starch differ in the position of their glucose links, and whereas cellulose is a straight-chain polymer of 150 to 300 units, starch has many branched chains containing anything from 70 to 400 units. Starch is found in roots, seeds and fruits of plants; for example corn, wheat, potatoes and rice. Cellulose is in all plant tissue; cotton is nearly 100% cellulose.

The reverse reaction, in which cellulose or starch react with water to break into many glucose molecules, is *hydrolysis*. The reaction and

fig 14.11 *the formation of protein*

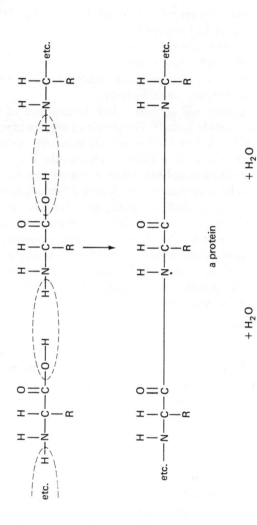

fig 14.12 *the formation of a glucose polymer*

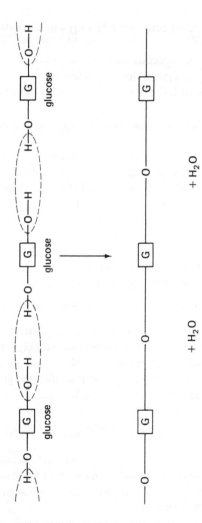

consequent breakdown of water with any substance is known as the *hydrolysis* of that substance.

The hydrolysis of starch occurs every time we eat a piece of bread. An organic catalyst, or *enzyme*, in saliva catalyses the hydrolysis of starch to glucose at body temperature. If a piece of bread is kept in the mouth for a few minutes, it begins to taste sweet as a result of this hydrolysis. The hydrolysis of starch can be carried out in the laboratory by gently boiling the polymer with a dilute acid, since $H^+(aq)$ will also catalyse the reaction. Starch can be identified by an intense blue–black colour which forms when a few drops of aqueous iodine are added. The disappearance of starch through its hydrolysis is demonstrated by:

(i) boiling starch solution for ten minutes with dilute hydrochloric acid in a test tube,

(ii) leaving some starch solution in a test tube with a few drops of saliva for one hour,

(iii) leaving a third test tube of starch solution untouched as a *control*.

Adding a few drops of iodine solution to each tube gives an intense blue colour only in tube (iii). In tubes (i) and (ii) starch has been hydrolysed.

The bodies of most animals cannot make enzymes to catalyse the hydrolysis of cellulose. The snail, which does enjoy a slow meal of wood, is an exception, and rabbits, horses, elephants, cows, etc., hold tiny micro-organisms in their digestive systems which are similarly able to hydrolyse cellulose from grass and leaves.

By 'burning' glucose, both plants and animals obtain energy:

$$C_6H_{12}O_6(s) + 6O_2(g) \longrightarrow 6CO_2(g) + 6H_2O(l) \quad \Delta H_r = -2826 \text{ kJ mol}^{-1}$$

Glucose for this purpose is generally stored in plants in the form of starch and in animals in the form of another glucose polymer called *glycogen*. Although both animals and plants are able to use up glucose to obtain energy, only plants can use energy to manufacture glucose. This is the process of *photosynthesis*. Carbon dioxide from the air, water from the soil and energy from the Sun's radiation are converted into glucose:

$$6CO_2(g) + 6H_2O(l) \xrightarrow{\text{sunlight}} C_6H_{12}O_6(s) + 6O_2(g)$$

The glucose energy cycle (Fig. 14.13) illustrates these conversions.

Of the glucose polymers, starch is most used by Man as a basic energy supply in the diet, whereas cellulose passes through the gut without reaction, acting as bulk roughage.

Industrially cellulose is processed by Man into *paper* and *rayon*.

(1) Paper

When wood (50% cellulose) is agitated with boiling alkali, only the fine fibres of cellulose remain undissolved. These are sieved out in a matted

EXPLAINING THE MOLE

Our assumption that matter is composed of individual atoms fits well with chemical observation. If atoms exist, then it should be possible to weigh them and to count them. However, since they are so small, we do not deal with individual atoms, but with 'standard packs' containing a fixed large number of atoms. This standard quantity is the *mole*.

15.1 RELATIVE ATOMIC MASS

Masses of different atoms can be compared using an expensive machine called a *mass spectrometer* (Fig. 15.1). Use is made of the observation that the path of moving charged particles is bent by a magnetic field.

In just the same way that it would require more force to change the

fig 15.1 *the mass spectrometer*

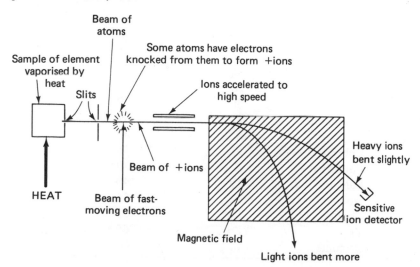

path of a double-decker bus than that of a ping-pong ball if both were passing by at 40 mph, so a greater magnetic field is needed in order to bend heavy ions into the ion detector than that required for the lighter ions. By measuring the magnetic field required to bend various ion beams into the ion detector, the masses of these singly charged ions, and thus the masses of the atoms from which they were derived, can be compared.

It has been agreed internationally that the masses of atoms should be stated relative to the mass of a particular atom of carbon, which is arbitrarily assigned 12 *atomic mass units* (amu). This means that hydrogen atoms, which happen to be twelve times lighter than carbon atoms, will have a mass of 1 amu. Other *relative atomic masses* are given in Table 15.1.

From Table 15.1 it can be seen that, for example, atoms of nitrogen are fourteen times heavier than atoms of hydrogen, and that atoms of magnesium are twenty-four times heavier than atoms of hydrogen.

Table 15.1   *relative atomic masses of some elements*

Element	Symbol	Approximate relative atomic mass
Hydrogen	H	1
Helium	He	4
Lithium	Li	7
Carbon	C	12
Nitrogen	N	14
Oxygen	O	16
Fluorine	F	19
Neon	Ne	20
Sodium	Na	23
Magnesium	Mg	24
Aluminium	Al	27
Silicon	Si	29
Phosphorus	P	31
Sulphur	S	32
Chlorine	Cl	35.5
Potassium	K	39
Calcium	Ca	40
Iron	Fe	56
Copper	Cu	63.5
Zinc	Zn	65.5
Bromine	Br	80

15.2 ISOTOPES

A pure sample of an element such as magnesium must contain only magnesium atoms. Yet the mass spectrometer consistently shows the presence of three different masses for atoms of this element. Magnesium atoms with a mass of 24 amu are most abundant, but small proportions of atoms with masses of 25 and 26 amu are also detected. Magnesium atoms must all contain 12 protons and 12 electrons (see Section 2.5), and therefore the different masses can only result from different numbers of neutrons. The mass of an electron is negligible compared with the mass of 1 amu for protons and neutrons (see Table 2.1). The composition of the three different magnesium atoms is determined in Table 15.2.

Table 15.2   *the three different magnesium isotopes*

Mass of magnesium atom	Number of protons (= number of electrons)	Therefore number of neutrons
24	12	12
25	12	13
26	12	14

Atoms with the same number of protons but different numbers of neutrons are called *isotopes.* Magnesium can be seen to have three isotopes The number of protons and neutrons together gives the mass of that individual atom, called the *mass number.* Just as atomic number (the number of protons) can be written as a subscript below the symbol for an atom, so the mass number is written as a superscript, as shown in Fig. 15.2. Thus the isotopes of magnesium are written as

fig 15.2 *positions for atomic and mass numbers before the symbol for an atom*

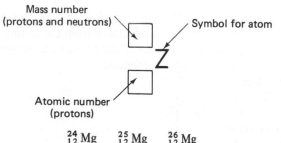

$$^{24}_{12}\text{Mg} \qquad ^{25}_{12}\text{Mg} \qquad ^{26}_{12}\text{Mg}$$

Because any sample of magnesium contains nearly 80% of the isotope with a mass number of 24, the relative atomic mass averages out to 24.3, and for most work a value of 24 is sufficiently accurate. All elements have isotopes, and in most cases one predominates. However, bromine has two isotopes of nearly equal abundance: $^{79}_{35}Br$ containing $79 - 35 = 44$ neutrons, and $^{81}_{35}Br$ with $81 - 35 = 46$ neutrons. The relative atomic mass of bromine therefore averages out to 80 amu.

15.3 RELATIVE MOLECULAR MASS

Comparison of the masses of atoms in Section 15.1 can be extended to molecules. Individual atoms of hydrogen and oxygen have relative masses of 1 and 16 amu (see Table 15.1), indicating that oxygen atoms are sixteen times heavier than hydrogen atoms. Water molecules, which contain

two atoms of hydrogen   and   one of oxygen,
will be (2×1) + (1×16) $= 18$

times heavier than an atom of hydrogen. Water molecules are said to have a *relative molecular mass* of 18 amu. Using Table 15.1 the relative molecular masses of other compounds can be determined in the same way. Four examples are given below.

(i) Sulphur dioxide,
   S O_2
   $32 + (16 \times 2) = 64$ amu.

(ii) Carbon dioxide,
   C O_2
   $12 + (16 \times 2) = 44$ amu.

(iii) Ammonia,
   N H_3
   $14 + (1 \times 3) = 17$ amu.

(iv) Ethanoic acid,
   C H_3 C O_2 H
   $12 + (1 \times 3) + 12 + (16 \times 2) + 1 = 60$ amu.

Ionic compounds do not form individual molecules, but nevertheless their relative molecular mass can be calculated from the formula in just the same manner as above for molecules. Relative formula masses are calculated below for a number of ionic compounds.

(i) Sodium chloride,
   Na   Cl
   $23 + 35.5 = 58.5$ amu.

(ii) Copper(II) oxide,
   Cu   O
   $63.5 + 16 = 79.5$ amu.

(iii) Aluminium bromide,

    Al Br_3

    27 + (80 × 3) = 267 amu.

(iv) Sodium hydroxide,

    Na O H

    23 + 16 + 1 = 40 amu.

13.4 AVOGADRO AND THE MOLE

Since in every light breath we inhale about forty thousand million million million atoms, the masses of individual atoms would be of little direct use in experimental work. However, whether comparing atoms singly, in tens or even in millions, their relative masses remain the same as long as we are consistent about the number of atoms or molecules chosen. For example, just as one atom of sulphur is 32 times heavier than one atom of hydrogen, so 20 million atoms of sulphur are also 32 times heavier than 20 million atoms of hydrogen. From research started by Lorenzo Avogadro (1776-1856), it can be calculated that

$$6 \times 10^{23} \text{ atoms of hydrogen weigh 1 g}$$

Hence

    6×10^{23} atoms of sulphur, being 32 times heavier, will weigh 32 g

    6×10^{23} atoms of carbon, being 12 times heavier, will weigh 12 g

    6×10^{23} molecules of water, being 16 + 2 = 18 times heavier, will weigh 18 g

In fact 6×10^{23} particles of any chemical weigh the relative atomic or molecular mass in grammes. 6×10^{23} is the *Avogadro number*. It has been chosen as a standard number of particles, and this number of particles of substance is *one mole* (abbreviated to *mol*) of that substance.

In any work involving chemical calculations it should be remembered that

One mole of substance contains the Avogadro number of particles

and

One mole of substance is the relative atomic or molecular mass of the substance in grammes

The following examples should help the reader to become really familiar with these two important statements.

Question 1

What is the mass of one mole of (a) carbon atoms, C; (b) oxygen atoms, O;

(c) water molecules, H_2O; (d) sodium hydroxide, NaOH; (e) hydrogen *atoms*, H; (f) hydrogen *molecules*, H_2?

(a) Carbon atoms, C:

relative atomic mass of carbon = 12 amu

Hence   1 mole of carbon atoms is   12 g

(b) Oxygen atoms, O:

relative atomic mass of oxygen = 16 amu

Hence   1 mole of oxygen atoms is   16 g

(c) Water molecules, H_2O:

relative molecular mass of H_2O = (2 + 16) = 18 amu

Hence   1 mole of water molecules is   18 g

(d) Sodium hydroxide, NaOH:

relative molecular mass of NaOH = (23 + 16 + 1) = 40 amu

Hence   1 mole of sodium hydroxide is   40 g

(e) Hydrogen *atoms*, H:

relative *atomic* mass of H = 1 amu

Hence   1 mole of hydrogen atoms is   1 g

(f) Hydrogen *molecules*, H_2:

relative *molecular* mass of H_2 = 2 amu

Hence   1 mole of hydrogen molecules is 2 g

Examples (e) and (f) show that it is important to specify the nature of the particles.

Question 2

What is the mass of (a) 3 moles of carbon dioxide molecules, CO_2; (b) 0.2 moles of calcium carbonate, $CaCO_3$; (c) 0.005 moles of sulphur atoms?

(a) 3 moles of carbon dioxide molecules, CO_2:

1 mole of CO_2 is   12+(16 × 2) =   44 g

Hence   3 moles of CO_2 is   44 × 3 =   132 g

(b) 0.2 moles of calcium carbonate, $CaCO_3$:

1 mole of $CaCO_3$ is   40+12+(16 × 3)= 100 g

Hence   0.2 moles of $CaCO_3$ is   100 × 0.2 =   20 g

(c) 0.005 moles of sulphur atoms, S:

1 mole of S is   32 g

Hence   0.005 moles of S is   32 × 0.005 =   0.16 g

Question 3

How many moles of atoms are in 8 g of oxygen?

	1 mole of O is	16 g
Hence	16 g of O atoms is	1 mole
	1 g of O atoms is	$\frac{1}{16}$ mole
	8 g of O atoms is	$\frac{1}{16} \times 8 = 0.5$ mole

Question 4

How many moles of molecules are in 3.4 g of ammonia?

	1 mole of NH_3 molecules is	$14 + 3 = 17$ g
Hence	17 g of NH_3 molecules is	1 mole
	1 g of NH_3 molecules is	$\frac{1}{17}$ mole
	3.4 g of NH_3 molecules is	$\frac{1}{17} \times 3.4 = 0.2$ mole

Question 5

How many moles are present in 10 g of sodium hydroxide?

	1 mole of NaOH is	$23+16+1 = 40$ g
Hence	40 g of NaOH is	1 mole
	1 g of NaOH is	$\frac{1}{40}$ mole
	10 g of NaOH is	$\frac{1}{40} \times 10 = 0.25$ mole

Question 6

How many atoms are in 4 moles of sulphur atoms?

	1 mole of anything contains	6×10^{23} particles
Hence	4 moles of sulphur atoms contain	$= 4 \times 6 \times 10^{23}$ atoms
		2.4×10^{24} atoms

Question 7

How many atoms are in 2 moles of sulphur dioxide molecules, SO_2?

	1 mole of anything contains	6×10^{23} particles
Hence	2 moles of SO_2 contain	$2 \times 6 \times 10^{23}$ molecules

Since each molecule contains 3 atoms (1 sulphur and 2 oxygen)

	2 moles of SO_2 contain	$3 \times 2 \times 6 \times 10^{23}$ atoms
		$= 3.6 \times 10^{24}$ atoms

Question 8

How many calcium ions and how many chloride ions are in 0.5 moles of calcium chloride, $CaCl_2$?

The formula $CaCl_2$ indicates that 1 mole of calcium chloride contains 1 mole of Ca^{2+} ions and 2 moles of Cl^- ions. Hence

0.5 moles of $CaCl_2$ contain   0.5 mole of Ca^{2+} ions
 and   1 mole of Cl^- ions
0.5 mole of Ca^{2+} ions is $0.5 \times 6 \times 10^{23}$ $= 3 \times 10^{23}$ Ca^{2+} ions
and   1 mole of Cl^- ions is $1 \times 6 \times 10^{23}$ $= 6 \times 10^{23}$ Cl^- ions

Section 6.2 showed that equations may be interpreted not only in terms of individual particles but also in terms of moles. The equation

$$2CO(g) + O_2(g) \longrightarrow 2CO_2(g)$$

tells us not only that

2 molecules of CO react with 1 molecule of O_2 to give 2 molecules of CO_2

but also that

2 *moles* of CO react with 1 *mole* of O_2 to give 2 *moles* of CO_2

Equations themselves are derived from experimental measurements of the reacting masses involved (see Section 16.3). Once determined, the equation is the starting point from which the mass of individual reactants and products may be determined.

Question 9

What mass of oxygen would be removed from the air by heating 10 g of powdered copper metal?
 Write equation

$$2Cu(s) + O_2(g) \longrightarrow 2CuO(s)$$

From equation

2 moles Cu react with 1 mole O_2

Hence

$2 \times 63.5 = 127$ g Cu react with $16 \times 2 = 32$ g O_2

1 g Cu reacts with $\frac{32}{127}$ g O_2

10 g Cu react with $\frac{32}{127} \times 10$ g O_2

2.5 g of oxygen would be removed

Question 10

If 100 g of washing soda crystals ($Na_2CO_3.10H_2O$) are heated, what mass of anhydrous sodium carbonate powder will remain?

Write equation

$$Na_2CO_3.10H_2O(s) \longrightarrow Na_2CO_3(s) + 10H_2O(l)$$

From equation

1 mole $Na_2CO_3.10H_2O$ gives 1 mole Na_2CO_3

Hence

$(23 \times 2) + 12 + (16 \times 3) + 10 \times (2+16)$	give	$(23 \times 2) + 12 + (16 \times 3)$
$= 286$ g $Na_2CO_3.10H_2O$		$= 106$ g Na_2CO_3
1 g $Na_2CO_3.10H_2O$	gives	$\frac{106}{286}$ g Na_2CO_3
100 g $Na_2CO_3.10H_2O$	give	$\frac{106}{286} \times 100$ g Na_2CO_3

37 g of anhydrous sodium carbonate would remain

Question 11

What mass of quicklime (calcium oxide) would be obtained when 1000 g of limestone (calcium carbonate) is heated?

Write equation

$$CaCO_3(s) \longrightarrow CaO(s) + CO_2(g)$$

From equation

1 mole of $CaCO_3$ produces 1 mole of CaO

Hence

$40 + 12 + (16 \times 3) = 100$ g $CaCO_3$ produce $40 + 16 = 56$ g CaO

1 g $CaCO_3$ produces $\frac{56}{100}$ g CaO

1000 g $CaCO_3$ produce $\frac{56}{100} \times 1000$ g CaO

560 g of quicklime would be produced by heating 1000 g limestone

Question 12

What mass of ammonia would have to be added to an excess of sulphuric acid in order to produce 1 kg of ammonium sulphate fertiliser?

Write equation

$$2NH_3(g) + H_2SO_4(aq) \longrightarrow (NH_4)_2SO_4(aq)$$

From equation

2 moles of NH_3 produce 1 mole of $(NH_4)_2SO_4$

Hence

$$2 \times (14 + 3) \qquad \text{produce} \qquad (14 + 4) \times 2 + 32 + (16 \times 4)$$
$$= 34 \text{ g NH}_3 \qquad \qquad = 132 \text{ g (NH}_4)_2\text{SO}_4$$

$$\tfrac{34}{132} \text{ g NH}_3 \qquad \text{produce} \qquad 1 \text{ g (NH}_4)_2\text{SO}_4$$

$$\tfrac{34}{132} \times 1000 \text{ g NH}_3 \qquad \text{produce} \qquad 1000 \text{ g (1 kg) (NH}_4)_2\text{SO}_4$$

257 g of ammonia would be required

15.5 THE BEHAVIOUR OF GASES

(a) Molar volume

In 1805 Joseph Gay-Lussac began to measure the volumes of gases reacting with one another. Today these experiments are simplified by the use of gas syringes. As an example, an experiment measuring the reacting volumes of nitrogen monoxide, NO, and oxygen, O_2, is outlined in Fig. 15.3. The two gases react to produce brown nitrogen dioxide, NO_2.

  (i) The three-way tap is turned to connect the inlet tube to syringe a.
 (ii) Syringe a is flushed with nitrogen monoxide several times before being filled with exactly 40 cm³ of this gas.
(iii) By this same procedure syringe b is filled with precisely 40 cm³ of oxygen.
 (iv) The tap is then turned to connect the two syringes, and the gases are pushed alternately from one syringe to the other until the brown colour of nitrogen dioxide becomes no darker.
  (v) With all the gases pushed into syringe a the total volume is measured, and found to be 60 cm³.

fig 15.3 *determining the reacting volumes of nitrogen monoxide and oxygen*

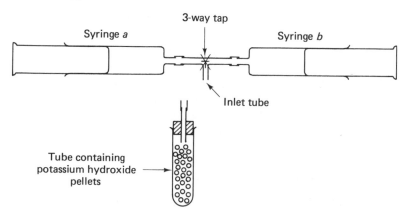

(vi) The inlet tube is connected to a sealed test tube containing potassium hydroxide pellets, and the tap turned to connect this to syringe a. The volume is seen to decrease steadily as the acidic nitrogen dioxide gas is absorbed into the solid alkali. When there is no further change of volume and the gas appears completely clear it may be assumed that all nitrogen dioxide has been removed. The residual gas is seen to occupy a volume of 20 cm^3.

(vii) Pumping this gas out over a glowing splint identifies it as oxygen, since the splint relights.

Summary

$$\text{nitrogen monoxide} + \text{oxygen} \longrightarrow \text{nitrogen dioxide}$$

At start   40 cm^3 40 cm^3 0 cm^3

The total volume left after reaction was 60 cm^3, of which 20 cm^3 was oxygen. Therefore $(60 - 20) = 40$ cm^3 is nitrogen dioxide.

Thus after reaction   0 cm^3 20 cm^3 40 cm^3

Therefore

40 cm^3 of NO reacted with   20 cm^3 of O_2 to form   40 cm^3 of NO_2

Thus

2 vol of NO reacted with   1 vol of O_2 to form   2 vol of NO_2

Similar experiments with other reactions show the volumes involved always to be in simple whole-number ratios. A comparison of these experimental results with the number of *moles* involved shows an interesting relationship, as the following examples show.

$$2NO(g) \; + \; O_2(g) \longrightarrow 2NO_2(g)$$
2 vol react with   1 vol to give   2 vol
2 mol react with   1 mol to give   2 mol

$$H_2(g) \; + \; Cl_2(g) \longrightarrow 2HCl(g)$$
1 vol reacts with   1 vol to give   2 vol
1 mol reacts with   1 mol to give   2 mol

$$N_2(g) \; + \; 3H_2(g) \longrightarrow 2NH_3(g)$$
1 vol reacts with   3 vol to give   2 vol
1 mol reacts with   3 mol to give   2 mol

The results show that, for example, three volumes of hydrogen contain three times as many moles as one volume of nitrogen, a quite different gas. Presumably only one volume of hydrogen would contain *the same* number of moles as one volume of nitrogen. Such arguments led Avogadro to conclude, in his theory of 1811, that equal volumes of any gas will contain the same number of molecules (and thus the same number of moles) provided only that the volumes are measured under identical conditions. Thus the space occupied by two million carbon dioxide molecules will be the same as the space taken up by two million hydrogen molecules, or that taken up by two million water vapour molecules, and so on. An equation will therefore not only tell us the ratio of moles reacting, but also the ratio of *gas volumes* reacting.

Question 13

How many cm^3 of oxygen will be used up when a bunsen burner completely burns $1000 \ cm^3$ of natural gas, CH_4?

Write equation

$$CH_4(g) + 2O_2(g) \longrightarrow CO_2(g) + 2H_2O(l)$$

From equation

1 vol of CH_4 reacts with 2 vol of O_2

Hence $1000 \ cm^3$ of CH_4 reacts with $2000 \ cm^3$ of O_2.

2000 cm^3 of oxygen will be used up

Question 14

In oxy-acetylene welding, what volume of oxygen will be used up for each cubic decimetre (dm^3) of ethyne (old name acetylene)? Also, what volume of carbon dioxide will be formed?

Write equation

$$2C_2H_2(g) + 5O_2(g) \longrightarrow 4CO_2(g) + 2H_2O(l)$$

From equation

2 vol of C_2H_2 react with 5 vol of O_2 to give 4 vol of CO_2

Hence $1 \ dm^3$ of C_2H_2 reacts with $2.5 \ dm^3$ of O_2 to give $2 \ dm^3$ of CO_2.

2.5 dm^3 of oxygen are used up, and 2 dm^3 of carbon dioxide
formed for each 1 dm^3 of ethyne burned

Experiments show that 1 mole of hydrogen molecules, H_2, occupy about $24 \ dm^3$ at room temperature and pressure (about 20°C and 1 atmosphere pressure). By Avogadro's theory this same volume of any other gas would also contain 1 mole. Thus

1 mole of any gas occupies 24 dm³ at room temperature and pressure

Question 15

5000 dm³ of carbon dioxide are absorbed by the leaves of an oak tree every day. How many moles and how many grammes of carbon dioxide is this?

$$24 \text{ dm}^3 \text{ of carbon dioxide is} \quad 1 \text{ mole}$$

Hence   1 dm³ of carbon dioxide is $\frac{1}{24}$ mole

   5000 dm³ of carbon dioxide is $\frac{1}{24}$ × 5000 mole

                                    = 208.3 mole

Also   1 mole of carbon dioxide
     has a mass of $12 + (16 \times 2) =$ 44 g

   208.3 mole of carbon dioxide
     have a mass of 44×208.3 =   9165 g

The tree would absorb 208.3 mole (i.e. 9165 g)
of carbon dioxide each day

Question 16

What is the mass of 1 dm³ of hydrogen gas, H_2 (measured at room temperature and pressure)?

   24 dm³ of hydrogen gas is   1 mole

                        and   1 mole of H_2 is 2 g

Hence   1 dm³ of hydrogen is $\frac{2}{24}$ g = 0.083 g

The mass of 1 dm³ of hydrogen gas at room temperature and
pressure is 0.083 g

Question 17

If 116 g of liquid butane, C_4H_{10}, are allowed to evaporate, what volume of butane vapour would be obtained at room temperature and pressure?

   1 mole of C_4H_{10} is $(12 \times 4) + (1 \times 10) = 58$ g

Hence   1 g of butane is $\frac{1}{58}$ moles

   116 g of butane is $\frac{1}{58} \times 116 = 2$ moles

Since 1 mole of any gas occupies 24 dm³ at room temperature and pressure,

$$2 \text{ moles of butane will occupy } 24 \times 2 \text{ dm}^3$$

$$= 48 \text{ dm}^3 \text{ under these conditions}$$

116 g of butane will occupy 48 dm³ at room temperature and pressure

Question 18

What volume of carbon dioxide would be obtained if 1 g of limestone (calcium carbonate) is heated?

Write equation

$$CaCO_3(s) \longrightarrow CaO(s) + CO_2(g)$$

From equation

1 mole of $CaCO_3$ gives 1 mole of CO_2

$$40 + 12 + (16 \times 3) = 100 \text{ g } CaCO_3 \text{ gives } 24 \text{ dm}^3 \text{ of } CO_2$$

Hence

$$1 \text{ g } CaCO_3 \text{ gives } \tfrac{24}{100} \text{ dm}^3 \text{ of } CO_2$$

$$= 0.24 \text{ dm}^3 \text{ of } CO_2$$

*0.24 dm³ or 240 cm³ of carbon dioxide would be produced
(measured at room temperature and pressure)*

(b) The gas laws

(i) *Pressure changes*

Imagine some gas, for example air, trapped in a gas syringe, as in Fig. 15.4. The pressure gauge in (a) shows the trapped air to be at atmospheric pressure. By pushing on the syringe plunger, pressure on this air can be increased to 2 and then 3 times that of the atmosphere. At the same time, the volume of trapped gas becomes smaller. As long as temperature remains constant during the experiment, doubling the pressure halves the volume, and so on. This was first noted by Robert Boyle in 1662. The relationship is expressed mathematically as

$$P = \text{a constant} \times \frac{1}{V}$$

(where P = pressure and V = volume of trapped gas). This is *Boyle's Law*.

fig 15.4 *the change of gas volume with pressure*

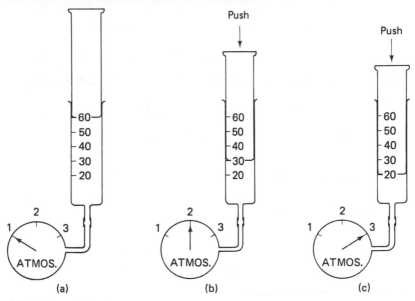

(a) (b) (c)

(ii) *Temperature changes*

The gas in the syringe can be heated evenly in a thermostatically controlled oven to show that the volume of a gas increases with temperature. Under constant pressure, experiment shows the increase in volume to be proportional to the increase in temperature. This relationship, discovered in 1787 by Jaques Charles, is *Charles' Law*.

A graph of temperature, in degrees centigrade, plotted against volume shows this proportionality by giving a straight line (Fig. 15.5). Furthermore, by continuing the straight line back to the temperature axis shows consistently for any gas that the volume would be zero at $-273°C$, and *less than zero* below this temperature. This led Lord Kelvin (1824-1907) to suggest that temperatures below $-273°C$ are impossible; that $-273°C$ is the *absolute zero* of temperature. Certainly, to date, temperatures below absolute zero have not been obtained. This being so, it would seem more sensible to make the absolute zero of temperature equal to zero degrees. Still using degrees centigrade but moving the zero gives the Kelvin absolute temperature scale (K) (Fig. 15.6).

Adding 273 to the temperature in centigrade gives the Kelvin absolute temperature. By using this temperature scale, Charles' law may be written mathematically as

$$T(K) = a \text{ constant} \times V$$

(where $T(K)$ is temperature in kelvins).

fig 15.5 *graph of volume of a trapped sample of gas against temperature*

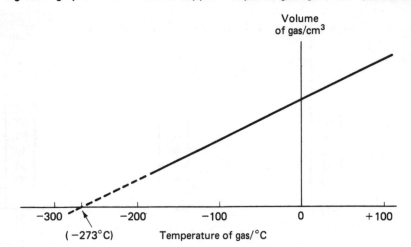

fig 15.6 *the Kelvin and centrigrade scales of temperature*

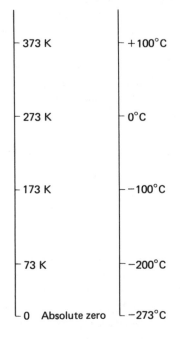

Boyle's and Charles' laws combine to give the expression

$$\frac{PV}{T(\text{K})} = \text{a constant for a fixed amount of gas}$$

This is useful in predicting the volume of a certain amount of gas if its pressure and/or its temperature were to be altered. Thus under conditions (1), where temperature is T_1, pressure is P_1 and volume is V_1, we can write

$$\frac{P_1 V_1}{T_1(K)} = \text{a constant}$$

and under conditions (2)

$$\frac{P_2 V_2}{T_2(K)} = \text{the same constant}$$

Hence

$$\frac{P_1 V_1}{T_1(K)} = \frac{P_2 V_2}{T_2(K)}$$

Question 19

A deep-sea diving bell holds $20\,000$ dm^3 of air at a temperature of $+7°C$ at a depth where the pressure is 10 atmospheres. What volume will this trapped air occupy when the bell is hauled to the surface where the temperature is $18°C$ and the pressure 1 atmosphere?

Let conditions (1) = underwater; conditions (2) = on surface:

$V_1 = 20\,000$ dm^3 $V_2 = ?$

$T_1(K) = (273 + 7) = 280$ K $T_2(K) = (273 + 18) = 291$ K

$P_1 = 10$ atmospheres $P_2 = 1$ atmosphere

$$\frac{P_1 V_1}{T_1(K)} = \frac{P_2 V_2}{T_2(K)}$$

Hence

$$V_2 = \frac{P_1 V_1 T_2(K)}{T_1(K) P_2} = \frac{10 \times 20\,000 \times 291}{280 \times 1} = 208\,000 \; dm^3$$

The trapped air will occupy 208 000 dm³ on the surface

This answer might explain why submariners have to keep breathing out when escaping from a submarine.

Question 20

1 mole of any gas is found by experiment to occupy precisely 24.04 dm^3 at $20°C$ and 1 atmosphere pressure. What volume would 1 mole of any gas occupy at $0°C$ and 1 atmosphere?

Let condition (1) = higher temperature; condition (2) = lower temperature:

$V_1 = 24.04$ dm^3 $V_2 = ?$

$T_1(K) = (273 + 20) = 293$ K $T_2(K) = 273$ K

$P_1 = 1$ atmosphere $P_2 = 1$ atmosphere

$$\frac{P_1 V_1}{T_1(K)} = \frac{P_2 V_2}{T_2(K)}$$

Hence

$$V_2 = \frac{P_1 V_1 T_2(K)}{T_1(K)P_2} = \frac{1 \times 24.04 \times 273}{293 \times 1} = 22.40 \text{ dm}^3$$

Thus 1 mole of any gas occupies 22.40 dm^3 at 0°C and 1 atmosphere

The condition of 0°C and 1 atmosphere pressure is frequently chosen as a standard, and is referred to as 's.t.p.'.

CHAPTER 16

APPLYING THE MOLE

In addition to serving as a tool with which to calculate formulae and equations from experimental results, the mole is chosen as the standard amount of substance against which to measure heat changes, reaction rate, concentration and so on.

16.1 DETERMINING FORMULAE

Formulae tell us the ratio of moles of atoms combined. The formula CO_2 indicates that, in carbon dioxide, for every one mole of carbon atoms there are two moles of oxygen atoms. The smallest whole-number ratio is the *empirical formula*. In the case of carbon dioxide, this is the same as the actual composition of the molecule, the *molecular formula*. Ethene, however, has the molecular formula C_2H_4 (this means that one molecule of ethene actually contains two carbon and four hydrogen atoms), but the empirical formula, giving only the simplest ratio of moles, is CH_2. Butane, with molecular formula C_4H_{10}, has empirical formula C_2H_5.

Empirical formulae are determined by the following steps.
(i) Experimentally determine the masses of each atom in the sample.
(ii) Convert this to the number of moles of each atom (divide by the relative atomic mass).
(iii) Adjust this ratio to simple whole numbers (divide each of the above by whichever is the smallest).
Examples of two experimental determinations of empirical formulae are given below.

(a) Copper(II) oxide
The reader should look at Fig. 16.1. The steps are as follows.
(i) The porcelain boat is weighed empty, and then containing powdered copper(II) oxide.
(ii) Hydrogen gas is blown gently through the apparatus. The gas is lit at

fig 16.1 *determination of the formula of copper oxide*

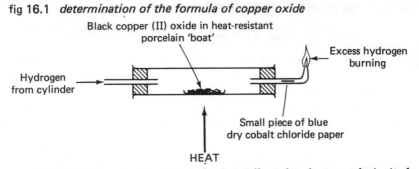

a jet only after some minutes, having collected and separately ignited a small test tube of the gas to check that it does not give a sudden explosion.

(iii) The black oxide is heated until it has all been reduced to salmon-pink copper:

$$CuO(s) + H_2(g) \longrightarrow Cu(s) + H_2O(l)$$

(iv) The boat should now contain just copper metal. After cooling, it is weighed. To check that reduction is complete, the boat could be replaced in the apparatus and the reduction process repeated. Re-weighing should show no further weight change.

Typical results

mass of boat alone	= 6.50 g
mass of boat + copper(II) oxide	= 7.78 g
mass of boat + copper after reduction	= 7.52 g

Hence   mass of copper = (7.52 − 6.50)   = 1.02 g
and   mass of oxygen combined = (7.78 − 7.52) = 0.26 g

	Copper	*Oxygen*
mass of each element	1.02 g	0.26 g
convert to moles (divide by relative atomic mass)	$\dfrac{1.02}{63.5}$ = 0.016 mol	$\dfrac{0.26}{16}$ = 0.016 mol
adjust ratio to whole numbers (divide by smallest)	$\dfrac{0.016}{0.016}$ = 1	$\dfrac{0.016}{0.016}$ = 1

The empirical formula is Cu_1O_1

(b) Lead iodide

(i) About 2 grammes of lead shot is weighed accurately into a large, narrow-necked conical flask.

(ii) 50 cm^3 of moderately concentrated nitric acid are added to this in a fume cupboard. A vigorous, complex reaction ensues and poisonous fumes of nitrogen dioxide pour off as lead is oxidised to soluble lead ions in the form of lead nitrate solution.

(iii) After diluting the solution with 50 cm^3 of distilled water, potassium iodide solution is added to give an intense yellow precipitate of lead iodide. The iodide solution is added, a little at a time, until further addition is seen to produce no more precipitate.

(iv) The precipitate is filtered off, washed with distilled water and then left in a desiccator for some days to dry thoroughly. The yellow powdered lead iodide is then weighed.

Typical results

	mass of lead shot	= 1.80 g
	mass of lead iodide obtained	= 4.00 g
Hence	mass of iodine combined	= 2.20 g

	Lead	Iodine
mass of each element	1.80 g	2.20 g
convert to moles (divide by relative atomic mass)	$\dfrac{1.80}{207}$ = 0.0087 mol	$\dfrac{2.20}{127}$ = 0.0173 mol
adjust ratio to simple whole numbers (divide by smallest)	$\dfrac{0.0087}{0.0087}$ = 1	$\dfrac{0.0173}{0.0087}$ = 1.99

The empirical formula is Pb_1I_2

Question 21

What is the empirical formula of a compound containing 52.0% zinc, 9.6% carbon and 38.4% oxygen by mass?

	Zinc	Carbon	Oxygen
mass of each element (per 100 g)	52.0 g	9.6 g	38.4 g
convert to moles (divide by relative atomic mass)	$\dfrac{52.0}{65}$ = 0.80 mol	$\dfrac{9.6}{12}$ = 0.80 mol	$\dfrac{38.4}{16}$ = 2.40 mol
adjust to simple whole numbers (divide by smallest)	$\dfrac{0.80}{0.80}$ = 1	$\dfrac{0.80}{0.80}$ = 1	$\dfrac{2.40}{0.80}$ = 3

The empirical formula is $Zn_1C_1O_3$

Question 22

3.94 g of hydrated copper(II) sulphate leave 2.52 g of the anhydrous salt after prolonged heating. What is the degree of hydration of this salt? The mass of combined water is 3.94 − 2.52 = 1.42 g.

	$CuSO_4$	H_2O
mass of each part	2.52 g	1.42 g
convert to moles (divide by relative molecular mass)	$\dfrac{2.52}{159.5}$ = 0.016 mol	$\dfrac{1.42}{18}$ = 0.079 mol
adjust ratio to simple whole numbers (divide by smallest)	$\dfrac{0.016}{0.016}$ = 1	$\dfrac{0.079}{0.016}$ = 4.94

The degree of hydration is 5 (i.e. formula is $CuSO_4.5H_2O$)

16.2 MOLARITY OF SOLUTIONS

Frequently it is useful to know the concentration of a solution in terms of the number of moles in a given volume.

The number of moles of substance dissolved in 1000 cm³ (1 dm³) of solution gives the molarity of that solution

Thus

a 2 molar (2 M) solution holds 2 moles in 1000 cm³ of solution
an X molar (X M) solution holds X moles in 1000 cm³ of solution

Question 23

How many moles are present in (a) 50 cm³ of 4 molar solution; (b) 25 cm³ of a 0.15 M solution?

(a) 50 cm³ of 4 molar (4 M) solution:

1000 cm³ of 4 M solution contains 4 moles (see above)

Hence   1 cm³ of 4 M solution contains $\dfrac{4}{1000}$ moles

50 cm³ of 4 M solution contains $\dfrac{4}{1000} \times 50$ moles

$$= 0.2 \text{ moles}$$

(b) 25 cm³ of a 0.15 molar (0.15 M) solution:

1000 cm³ of 0.15 M solution contains   0.15 moles

Hence   1 cm³ of 0.15 M solution contains $\dfrac{0.15}{1000}$ moles

25 cm³ of 0.15 M solution contains $\dfrac{0.15}{1000} \times 25$ moles

$$= \quad 0.00375 \text{ moles}$$

Question 24

What is the molarity of a solution containing 0.05 moles in 80 cm³?

80 cm³ contain   0.05 moles

Hence   1 cm³ contains $\dfrac{0.05}{80}$ moles

1000 cm³ contains $\dfrac{0.05}{80} \times 1000$ moles $= 0.625$ moles

The solution is 0.625 M

Question 25

What is the molarity of a solution containing 4 g of sodium hydroxide in 450 cm^3 of solution?

$$1 \text{ mole of NaOH is } (23 + 16 + 1) = 40 \text{ g}$$

Hence $4 \text{ g of NaOH is } \dfrac{4}{40} \text{ mole} = 0.1 \text{ mole}$

$$450 \text{ cm}^3 \text{ contains } 0.1 \text{ mole}$$

Hence $1 \text{ cm}^3 \text{ contains } \dfrac{0.1}{450} \text{ moles}$

$$1000 \text{ cm}^3 \text{ contains } \dfrac{0.1}{450} \times 1000 \text{ moles} = 0.22 \text{ moles}$$

The solution is 0.22 molar (0.22 M)

Question 26

What mass of silver nitrate is contained in 40 cm^3 of a 0.05 molar solution of this salt?

$$1000 \text{ cm}^3 \text{ contain } 0.05 \text{ mole}$$

Hence $1 \text{ cm}^3 \text{ contains } \dfrac{0.05}{1000} \text{ mole}$

$$40 \text{ cm}^3 \text{ contains } \dfrac{0.05}{1000} \times 40 \text{ mole} = 0.002 \text{ mole}$$

$$1 \text{ mole of AgNO}_3 = 108 + 14 + (16 \times 3) = 170 \text{ g}$$

Therefore

$$0.002 \text{ mole} = (170 \times 0.002) = 0.34 \text{ g}$$

40 cm^3 of 0.05 M AgNO$_3$(aq) contains 0.34 g

16.3 DETERMINING EQUATIONS

In addition to knowing the formulae of the chemicals involved, before the equation for a particular reaction can be written the reacting proportions of these chemicals must also be determined. The following experiments illustrate how the relative numbers of moles reacting together may be found.

(a) The complete neutralisation of sodium hydroxide by sulphuric acid
Assuming that the formulae of the reactants are already known, the purpose of this experiment is to find the relative number of moles of acid and alkali which react. The neutralisation point can be determined using an indicator, or by measuring the temperature change.

(i) With indicator

Details of titrating acid with alkali were given in Section 9.7. Exactly 1.0 M sulphuric acid is used in the burette. Exactly 25.0 cm³ of 0.8 M sodium hydroxide solution is pipetted into the conical flask, and two drops of methyl orange indicator added. Acid is run slowly into the flask with constant swirling of the contents, until the yellow alkali colour of the indicator has just turned permanently orange. The titration is repeated until consecutive results agree.

Typical results

$$\text{volume of 0.8 M NaOH(aq) used } = 25.0 \text{ cm}^3$$

$$\text{volume of 1.0 M H}_2\text{SO}_4\text{(aq) used } = 10.0 \text{ cm}^3$$

Determine number of moles of alkali used

$$1000 \text{ cm}^3 \text{ of 0.8 M solution contain } 0.8 \text{ mole}$$

Hence $\quad\quad$ 1 cm^3 of 0.8 M solution contain $\dfrac{0.8}{1000}$ mole

$$25 \text{ cm}^3 \text{ of 0.8 M solution contain } \dfrac{0.8}{1000} \times 25 \text{ mole}$$

$$= 0.02 \text{ moles of NaOH}$$

Determine number of moles of acid used

$$1000 \text{ cm}^3 \text{ of 1.0 M solution contain } 1.0 \text{ mole}$$

Hence $\quad\quad$ 1 cm^3 of 1.0 M solution contain $\dfrac{1.0}{1000}$ mole

$$10 \text{ cm}^3 \text{ of 1.0 M solution contain } \dfrac{1.0}{1000} \times 10 \text{ mole}$$

$$= 0.01 \text{ moles of H}_2\text{SO}_4$$

Thus 0.02 moles of NaOH react with 0.01 moles of H_2SO_4, indicating that the equation involves

2 moles of NaOH reacting with 1 mole of H_2SO_4

(ii) *By temperature change*

In this experiment, acid from a burette is run into the alkali in a polystyrene beaker. Exactly 50 cm³ of the 0.8 M sodium hydroxide solution are measured by pipette into the polystyrene beaker. Polystyrene is a good heat insulator, so little of the heat given out by the reaction is lost from the beaker. The temperature of this solution is recorded using a thermometer. Then, stirring thoroughly with the thermometer, 1.0 M sulphuric acid is added from the burette, and the temperature recorded after each 2 cm³ addition. Results are plotted on a graph of temperature against volume of acid used. Because the reaction is exothermic, the temperature rises steadily as more acid is added until the point is reached when all the alkali has been neutralised. Further addition of sulphuric acid will then simply cool down the hot solution.

Typical results

A typical set of results is shown in Fig. 16.2. From the graph

fig 16.2 *graph of temperature of reaction mixture against volume of acid added*

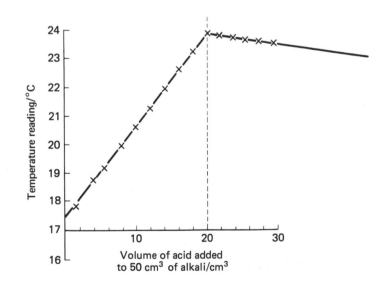

$$50 \text{ cm}^3 \text{ of } 0.8 \text{ M NaOH(aq) was neutralised by}$$

$$20 \text{ cm}^3 \text{ of } 1.0 \text{ M H}_2\text{SO}_4\text{(aq)}$$

Determine number of moles of alkali used

$$1000 \text{ cm}^3 \text{ of } 0.8 \text{ M solution contain } 0.8 \text{ mole}$$

Hence $50 \text{ cm}^3 \text{ of } 0.8 \text{ M solution contain } \dfrac{0.8}{1000} \times 50 \text{ mole}$

$$= 0.04 \text{ mole}$$

Determine number of moles of acid used

$$1000 \text{ cm}^3 \text{ of } 1.0 \text{ M solution contain } 1.0 \text{ mole}$$

Hence $20 \text{ cm}^3 \text{ of } 1.0 \text{ M solution contain } \dfrac{1.0}{1000} \times 20 \text{ mole}$

$$= 0.02 \text{ mole}$$

Thus 0.04 moles of NaOH(aq) were neutralised by 0.02 moles of H_2SO_4(aq), confirming the ratio calculated from the previous experiment.

(b) The reaction between lead nitrate and potassium iodide solutions
Lead ions and iodide ions combine to form a dense yellow precipitate of lead iodide.

Exactly 1.0 M lead nitrate and 1.0 M potassium iodide solution are separately measured from burettes into nine similar test tubes according to the volumes specified below.

tube number	1	2	3	4	5	6	7	8	9	
vol. of 1.0 M $Pb(NO_3)_2$(aq)	1	2	3	4	5	6	7	8	9	cm³
vol. of 1.0 M KI(aq)	9	8	7	6	5	4	3	2	1	cm³

After stirring each mixture, the resulting yellow precipitates are left to stand for ten minutes before the height of precipitate in each tube is measured. A graph of precipitate height against composition of mixture is then plotted (Fig. 16.3). The greatest precipitate height will correspond to the mixture in which both solutions reacted completely. From the graph

$$3.3 \text{ cm}^3 \text{ of } 1.0 \text{ M Pb(NO}_3)_2\text{(aq) would react completely with}$$

$$6.7 \text{ cm}^3 \text{ of } 1.0 \text{ M KI(aq)}$$

288

fig 16.3 *graph of precipitate height against composition of the mixture*

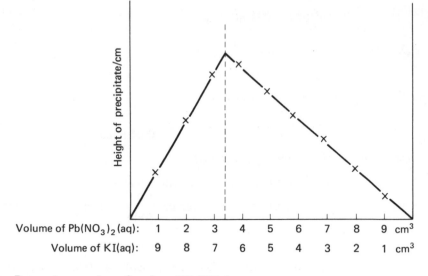

Volume of Pb(NO₃)₂(aq): 1 2 3 4 5 6 7 8 9 cm³
Volume of KI(aq): 9 8 7 6 5 4 3 2 1 cm³

Determine number of moles of Pb(NO₃)₂

$$1000 \text{ cm}^3 \text{ of } 1.0 \text{ M solution contain } 1.0 \text{ mole}$$

Hence 3.3 cm³ of 1.0 M solution contain $\dfrac{1.0}{1000} \times 3.3$ mole

$$= 0.0033 \text{ mole}$$

Determine number of moles of KI

$$1000 \text{ cm}^3 \text{ of } 1.0 \text{ M solution contain } 1.0 \text{ mole}$$

Hence 6.7 cm³ of 1.0 M solution contain $\dfrac{1.0}{1000} \times 6.7$ mole

$$= 0.0067 \text{ mole}$$

Thus 0.0033 moles of Pb(NO₃)₂ reacted with 0.0067 moles of KI, indicating that

1 mole of Pb(NO₃)₂ would react with 2 moles of KI

16.4 DETERMINING RELATIVE MOLECULAR MASS

The relative molecular mass of any gas or vapour can be determined by applying the statement that

1 mole of any gas occupies 24 dm³ at room temperature and pressure
(see Section 15.5)

Question 27

105 cm^3 of a gaseous oxide of sulphur weigh 0.28 g. What is the relative molecular mass of this oxide?

$$105 \text{ cm}^3 \text{ weigh } 0.28 \text{ g}$$

Hence $\qquad 1 \text{ cm}^3 \text{ weigh } \dfrac{0.28}{105} \text{ g}$

$$24\,000 \text{ cm}^3 \text{ weigh } \dfrac{0.28}{105} \times 24\,000 = 64 \text{ g}$$

Also $\qquad 24\,000 \text{ cm}^3$ of a gas is 1 mole

Thus $\qquad$ 1 mole of this gas is 64 g

The relative molecular mass is 64

Question 28

0.20 g of a hydrocarbon with empirical formula C_3H_7 occupies 55.8 cm^3 at room temperature and pressure. Deduce its relative molecular mass and hence its molecular formula.

$$55.8 \text{ cm}^3 \text{ weigh } 0.20 \text{ g}$$

Hence $\qquad 1 \text{ cm}^3 \text{ weighs } \dfrac{0.20}{55.8} \text{ g}$

$$24\,000 \text{ cm}^3 \text{ weigh } \dfrac{0.20}{55.8} \times 24\,000 \text{ g} = 86 \text{ g}$$

Thus the relative molecular mass of the hydrocarbon is 86. The empirical formula is given as C_3H_7, which has a molecular mass of $(12 \times 3) + (1 \times 7) = 43$.

Since actual molecular mass is 86 (i.e. twice 43), molecular formula must be twice the empirical formula.

Therefore the molecular formula is C_6H_{14}

Relative molecular mass may also be determined using the mass spectrometer and by measurements on solutions, but these methods are beyond the scope of this book.

16.5 **THE MOLE IN ELECTROLYSIS**

A re-read of Chapter 8 would provide a worthwhile introduction to this section.

(a) The Faraday

The equation

$$Al^{3+} + 3e \longrightarrow Al$$

can be interpreted on the atomic scale as meaning

1 aluminium ion is converted by 3 electrons into 1 aluminium atom

or on the larger scale as meaning

1 *mole* of aluminium ions is converted by 3 *moles* of electrons
into 1 *mole* of aluminium atoms

Moles of atoms, molecules or ions can be measured out through weighing, but moles of electrons correspond to a certain amount of electricity.

1 mole of electron is 1 faraday (1 F) of electricity

Thus it can be seen that 3 faradays of electricity are required to deposit 1 mole of aluminium atoms.

It can be seen from Table 16.1 that

1 faraday is the quantity of electricity required to deposit
1 mole of singly charged ions,
$\frac{1}{2}$ mole of doubly charged ions,
and $\frac{1}{3}$ mole of triply charged ions.

Table 16.1   *the relationship between the ionic charge and amount of substance deposited by 1 faraday*

Equation at electrode	Equation adjusted for 1 electron	Amount deposited by 1 faraday
$Ag^+ + 1e \longrightarrow Ag$	$Ag^+ + 1e \longrightarrow Ag$	1 mole of Ag atoms
$K^+ + 1e \longrightarrow K$	$K^+ + 1e \longrightarrow K$	1 mole of K atoms
$Cu^{2+} + 2e \longrightarrow Cu$	$\frac{1}{2}Cu^{2+} + 1e \longrightarrow \frac{1}{2}Cu$	$\frac{1}{2}$ mole of Cu atoms
$Mg^{2+} + 2e \longrightarrow Mg$	$\frac{1}{2}Mg^{2+} + 1e \longrightarrow \frac{1}{2}Mg$	$\frac{1}{2}$ mole of Mg atoms
$Al^{3+} + 3e \longrightarrow Al$	$\frac{1}{3}Al^{3+} + 1e \longrightarrow \frac{1}{3}Al$	$\frac{1}{3}$ mole of Al atoms
$Cl^- \longrightarrow \frac{1}{2}Cl_2 + 1e$	$Cl^- \longrightarrow \frac{1}{2}Cl_2 + 1e$	1 mole of Cl atoms ($\frac{1}{2}$ mole of Cl_2 molecules)

Question 29

What weight of copper would be deposited from the electrolysis of copper(II) sulphate solution by 0.3 faradays?

Write equation

$$Cu^{2+}(aq) + 2e \longrightarrow Cu(s)$$

From equation

2 faradays deposit   1 mole of Cu atoms

$$= \quad 63.5 \text{ g of copper}$$

Hence   1 faraday deposits $\dfrac{63.5}{2}$ g of copper

0.3 faradays deposit $\dfrac{63.5}{2} \times 0.3$ g of copper

$$= \quad \textit{9.52 g of copper}$$

Question 30

What volume of oxygen gas (at room temperature and pressure) would be liberated from the anode in the electrolysis of aqueous sodium hydroxide solution by 0.5 faradays?

Write equation
at anode

$$4OH^-(aq) \longrightarrow 2H_2O(l) + O_2(g) + 4e$$

From equation

4 faradays liberate   1 mole of O_2 molecules

$$= \quad 24 \text{ dm}^3 \text{ of gas}$$

Hence   1 faraday liberates $\dfrac{24}{4}$ dm^3

0.5 faradays liberate $\dfrac{24}{4} \times 0.5$ dm^3

$$= \quad \textit{3 dm}^3 \textit{ or 3000 cm}^3 \textit{ of gas}$$

Question 31

How many faradays will be required to produce 1 kg of aluminium by electrolysis of molten aluminium oxide?

Write equation
at cathode

$$Al^{3+}(l) + 3e \longrightarrow Al(s)$$

From equation

1 mole of Al atoms is deposited by 3 faradays
27 g of aluminium is deposited by 3 faradays

Hence

1 g of aluminium is deposited by $\dfrac{3}{27}$ faradays

Therefore

1 kg or 1000 g of aluminium is deposited by $\dfrac{3}{27} \times 1000$ faradays

$$= 111.1 \text{ faradays}$$

111.1 faradays would be required to obtain 1 kg of aluminium

(b) Electrolysis in series

In the circuit shown in Fig. 16.4, the same quantity of electricity flows through each voltameter. This is so because the same number of electrons that are pushed out by the negative terminal of the cell are pulled in by the positive terminal. When connected in this way, the voltameters are said to be wired *in series*.

fig 16.4 *electrolysis in series*

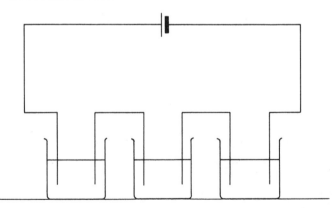

Question 32

An electric current is passed through two voltameters connected in series. The first contains dilute aqueous copper(II) sulphate and the second aqueous silver nitrate. If 2.20 g of copper are deposited on the cathode of the first voltameter, what mass of silver will collect on the cathode of the second?

Write equation at
cathode of first

$$Cu^{2+}(aq) + 2e \longrightarrow Cu(s)$$

From equation

$$\left\{ \begin{array}{l} \text{1 mole of Cu atoms} \\ \text{63.5 g of copper} \end{array} \right\} \text{ are deposited by 2 faradays}$$

Hence

$$\text{2.2 g of copper is deposited by } \frac{2}{63.5} \times 2.20 \text{ faradays}$$

$$= 0.069 \text{ faradays}$$

0.069 faradays must flow through the second voltameter.

Write equation at
cathode of second

$$Ag^{+}(aq) + 1e \longrightarrow Ag(s)$$

From equation

1 faraday deposits   1 mole of Ag atoms
$= $ 108 g of silver

Hence

0.069 faradays deposit $(108 \times 0.069) = 7.45$ g of silver

7.45 g of silver will be deposited at the cathode of the second voltameter

Question 33

Electricity is passed through two voltameters connected in series. 200 cm^3 of oxygen (at room temperature and pressure) is collected from the anode of the first, which contains dilute sulphuric acid. What mass of copper will collect on the cathode of the second voltameter, containing copper(II) sulphate solution?

Write equation
at anode of first

$$4OH^{-}(aq) \longrightarrow 2H_2O(l) + O_2(g) + 4e$$

From equation

1 mole of O_2 molecules is deposited by 4 faradays

24 dm^3 of oxygen is deposited by 4 faradays

Hence

1 cm^3 of oxygen is deposited by $\dfrac{4}{24000}$ faradays

200 cm^3 of oxygen is deposited by $\dfrac{4}{24000} \times 200$ faradays

$= 0.033$ faradays

0.033 faradays must also flow through the second voltameter.

Write equation at
cathode of second

$$Cu^{2+}(aq) + 2e \longrightarrow Cu(s)$$

From equation

2 faradays deposit   1 mole of Cu atoms

$=$ 63.5 g of copper

Hence   1 faraday deposits $\dfrac{63.5}{2}$ g

0.033 faradays deposit $\dfrac{63.5}{2} \times 0.033$ g

$=$ 1.05 g of copper

1.05 of copper will be deposited in the second voltameter

(c) Counting faradays

Meters measuring the number of faradays of electricity are few and far between. However ammeters, measuring electrical flow rate in amps, and clocks are common enough, and these can be used to measure faradays in a slightly less direct manner.

Most people who use electric currents care little for moles, and choose to measure quantities of electricity in *coulombs*.

1 coulomb of electricity passes when a current of 1 amp
flows for 1 second

Thus

$$\text{amps} \times \text{seconds} = \text{coulombs}$$

Coulombs are related to faradays by the ungainly figure of 96 500:

$$96\,500 \text{ coulombs} = 1 \text{ faraday}$$

Question 34

A bedside light bulb uses a current of 0.16 amps. How many faradays will be used by this light each hour?

$$\text{amps} \times \text{seconds} = \text{coulombs}$$

and

$$1 \text{ hour} = 60 \times 60 \text{ seconds}$$

Hence

$$\text{quantity of electricity used} = 0.16 \times 60 \times 60 \text{ coulombs}$$

$$= 576 \text{ coulombs}$$

Now

$$96\,500 \text{ coulombs} = 1 \text{ faraday (see above)}$$

Therefore

$$1 \text{ coulomb} = \frac{1}{96500} \text{ faradays}$$

$$576 \text{ coulombs} = \frac{1}{96500} \times 576 \text{ faradays}$$

$$= 0.006 \text{ faradays}$$

0.006 faradays will be used by the light each hour

Question 35

What mass of silver can be reclaimed by electrolysing an aqueous solution of silver nitrate with a current of 5 amps flowing for 2 hours?

$$\text{amps} \times \text{seconds} = \text{coulombs}$$

Hence

$$\text{quantity of electricity} = 5 \times 2 \times 60 \times 60 \text{ coulombs}$$

$$= 36000 \text{ coulombs}$$

$$= \frac{36000}{96500} \text{ Faradays}$$

$$= 0.37 \text{ Faradays}$$

*Write equation
at cathode*

$$Ag^+(aq) + 1e \longrightarrow Ag(s)$$

From equation

> 1 faraday deposits   1 mole of Ag atoms
>
> $=$ 108 g of silver

Therefore

> 0.37 faradays deposit 108×0.37 g
>
> *40 g of silver will be reclaimed*

Question 36

The large-scale industrial production of aluminium uses a current of 25 000 amps in the electrolysis of molten aluminium oxide. What mass of aluminium would you expect to be produced each hour?

$$\text{amps} \times \text{seconds} = \text{coulombs}$$

Hence

> quantity of electricity in one hour $= 25\,000 \times 60 \times 60$ coulombs
>
> $= 90\,000\,000$ coulombs
>
> $= \dfrac{90\,000\,000}{96\,500}$ faradays
>
> $= 933$ faradays

*Write equation
at cathode*

$$Al^{3+}(l) + 3e \longrightarrow Al(s)$$

From equation

> 3 faradays deposit   1 mole of Al atoms
>
> $=$ 27 g of aluminium

Therefore   933 faradays deposit $\dfrac{27}{3} \times 933$ g

> $=$ 8397 g
>
> $=$ approximately 8.4 kg

8.4 kg of aluminium would be deposited each hour

Question 37

A current of 1.02 amps is passed through a voltameter containing an alkaline solution of a copper compound. After precisely 1 hour, 2.42 g of copper had been deposited on the cathode. Determine the magnitude of the positive charge on the copper ions in the solution.

$$\text{amps} \times \text{seconds} = \text{coulombs}$$

Hence

$$\text{quantity of electricity} = 1.02 \times 60 \times 60 \text{ coulombs}$$

$$= 3672 \text{ coulombs}$$

$$= \frac{3672}{96500} \text{ faradays}$$

$$= 0.038 \text{ faradays}$$

Thus 0.038 faradays deposited 2.42 g of copper. Hence

$$1 \text{ faraday would deposit} \quad \frac{2.42}{0.038} \text{ g}$$

$$= 63.7 \text{ g of copper}$$

$$= \frac{63.7}{63.5} \text{ moles of Cu}$$

$$= 1 \text{ mole of Cu}$$

1 faraday deposits 1 mole of copper atoms

suggesting the ions in the electrolysis to be Cu^{1+}:

$$Cu^+(aq) + 1e \longrightarrow Cu(s)$$

16.6 THE MOLE IN THERMOCHEMISTRY

A re-read of Chapter 5 would provide a worthwhile introduction to this section.

When comparing energy changes resulting from the interaction of particles, it would seem logical to consider the same number of particles in each comparison. Energy changes are conventionally considered *per mole* of chemical reacting.

(a) The joule

The unit of energy is the joule. Being such a small unit (1 joule would keep a reading lamp alight for 0.02 second), thousands of joules, or kilojoules (kJ), are more frequently used. In simple experiments, energy changes can be measured conveniently by making a reaction heat up or cool down a known mass of water. 4.2 joules of energy are required to heat 1 g of water by 1°C. Conversely 4.2 joules of energy have to be lost in order to cool 1 g of water by 1°C. This is the meaning of the statement that 'the *specific heat* of water is 4.2 J g^{-1} °C^{-1}.

If the mass of water used in an experiment is known and the temperature change measured, then

Number of joules = 4.2 × mass of water × temperature change of water

If the water became *hotter*, then energy must have been given up by the reaction and ΔH would be *negative*. If the water became *cooler*, then energy must have been taken in by the reaction, requiring ΔH to be *positive*.

(b) Experimental measurement of ΔH values
The following experiments illustrate how ΔH values may be obtained in practice.

(i) *Determination of* $\Delta H_{solution}$ *for potassium iodide*

100 cm³ of distilled water is transferred by measuring cylinder to a heat-insulating polystyrene beaker. The water is stirred with a thermometer and its temperature recorded. 16.6 g of potassium iodide, ready weighed out on a watchglass, is quickly poured into the water and the mixture stirred continually. The water temperature drops steadily, and the minimum temperature is noted.

Typical results

mass of potassium iodide	= 16.6 g
volume of water	= 100 cm³
initial temperature of water	= 18.2°C
minimum temperature of water after adding potassium iodide	= 11.0°C

100 cm³ of water is approximately 100 g of water (density = 1 g cm⁻³).
Fall in temperature of water = (18.2 − 11.0) = 7.2°C. Hence

$$energy\ change = 4.2 \times 100 \times 7.2\ joules$$

$$= 3024\ joules$$

1 mole of KI = (39 + 127) = 166 g. Therefore

$$16.6\ g\ of\ KI\ is\ \frac{16.6}{166} = 0.1\ mole$$

0.1 mole of KI took 3024 joules of energy to dissolve. Therefore

1 mole of KI would take $\dfrac{3024}{0.1}$ = 30240 joules of energy to dissolve

Since energy is *taken from* the water the reaction is *endothermic*, and ΔH will be *positive*.

$$\Delta H_s(KI) = + 30.24 \text{ kJ mol}^{-1}$$

(ii) *Determination of* $\Delta H_{reaction}$ *for* $Zn(s) + Cu^{2+}(aq) \rightarrow Zn^{2+}(aq) + Cu(s)$

Using a measuring cylinder, 100 cm³ of 1.0 M copper(II) sulphate solution is transferred to a heat-insulating polystyrene beaker. The solution is stirred with a thermometer and its temperature recorded.

100 cm³ of this 1.0 M solution contains $\dfrac{1}{1000} \times 100 = 0.1$ moles

The equation indicates that 0.1 moles of Cu^{2+} ions will react with 0.1 moles of Zn atoms. A minimum of 0.1 moles of zinc atoms must be weighed out if all the copper(II) ions are to react. Thus *at least* (0.1 × 65.5) = 6.55 g of powdered zinc are weighed onto a watchglass and poured quickly into the insulated beaker. The mixture is stirred continually; the temperature rises rapidly and the maximum temperature is noted.

Typical results

	volume of 1.0 M CuSO₄(aq)	= 100 cm³
	initial temperature of solution	= 18.6°C
	maximum temperature reached by mixture	= 68.8°C
Thus	rise in temperature = (68.8 − 18.6)	= 50.2°C

Therefore

energy change = 4.2 × 100 × 50.2 joules
= 21 084 joules
= 21 kJ (approximately)

number of moles of CuSO₄ in 100 cm³ of 1.0 M solution =

$$\dfrac{1}{1000} \times 100 = 0.1 \text{ moles}$$

0.1 mole, in reacting, gave 21 kJ. Hence

1 mole would give $\dfrac{21}{0.1}$ = 210 kJ

Since energy was *given to* the water, the reaction is *exothermic* and ΔH *negative*.

Thus, for the reaction Zn(s) + Cu²⁺(aq)→Zn²⁺(aq) + Cu(s),

$$\Delta H_r = -210 \ kJ \ mol^{-1}$$

(iii) *Determination of* $\Delta H_{combustion}$ *for ethanol,* C_2H_5OH

A small spirit burner is rinsed out and topped up with ethanol. It is dried thoroughly on the outside and, with the glass cover in position to prevent evaporation, it is weighed accurately.

Using a measuring cylinder, 400 cm³ of water is transferred to a conical flask which is fixed above the burner as shown in Fig. 16.5. The water is stirred gently with a thermometer and its temperature recorded. The burner is then lit and the water stirred until a rise of, for example, 40°C has been recorded. The flame is then immediately snuffed out with the glass cover, and the burner reweighed with its cover.

Typical results

volume of water in flask	= 400 cm³
initial temperature of water	= 18.0°C
final temperature of water	= 58.0°C
initial mass of ethanol + burner	= 62.41 g
mass of remaining ethanol + burner after experiment	= 60.13 g

fig 16.5 *determining* ΔH_c

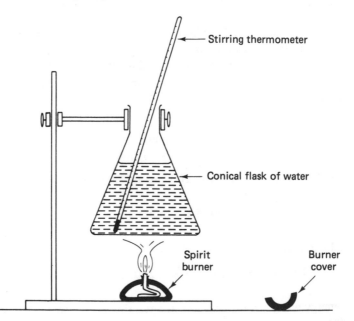

Stirring thermometer

Conical flask of water

Spirit burner

Burner cover

Therefore

temperature rise of water = $(58 - 18)$ = $40°C$

mass of ethanol burned =

$(62.41 - 60.13)$ = $2.28\,g$

$400\,cm^3$ of water is approximately 400 g of water, and

heat given to the water = $4.2 \times 400 \times 40$ joules

= 67 200 joules

= 67 kJ (approximately)

1 mole of ethanol (C_2H_5OH) is $(12 \times 2) + (1 \times 5) + 16 + 1 = 46$ g.

mass of ethanol burned = $2.28\,g$ = $\dfrac{2.28}{46}$ mole

= 0.05 mole

0.05 mole of ethanol burned to give 67 kJ. Therefore

1 mole of ethanol would burn to give $\dfrac{67}{0.05}$ kJ

= 1340 kJ

Since the water temperature increased, heat was *given out* by the reaction, showing it to be *exothermic*, and ΔH to be negative. Thus

$$\Delta H_c(C_2H_5OH) = -1340\ kJ\ mol^{-1}$$

QUESTIONS

CHAPTER 1

1 Do the following possess kinetic energy, potential energy or both?
   (a) A bullet travelling through the air.
   (b) A clockwork musical box when it is ready to be played.
   (c) Water high in a mountain lake.
   (d) A falling apple just before it hits the ground.
   (e) A swinging pendulum.
   (f) A positive and negative charge forcibly held apart.
   (g) The moon travelling around the Earth.

2 By what means is energy transferred in the following two examples?
   (a) Heat from a log fire burning our faces.
   (b) Heat from a cooking pan being carried along to its handle.

3 Suggest a possible reason for some substances, such as iron, conducting heat better than others, such as concrete.

4 Invent experiments using everyday materials to demonstrate each of the following:
   (a) the conversion of potential energy into kinetic energy;
   (b) the conversion of potential energy into heat;
   (c) the conversion of kinetic energy into potential energy.
   Draw a labelled diagram in each case.

CHAPTER 2

1 Draw out the atomic structures of each of the following:
   (a) $_{20}Ca$,   (b) $_8O$,   (c) $_{18}Ar$,
   (d) $_{16}S$,   (e) $_4Be$,   (f) $_2He$.

2 Which of the elements in question 1 would you expect to be
(a) very stable,
(b) in group 2 of the Periodic Table,
(c) in group 6 of the Periodic Table?
Give a reason for each answer.

3 An atom has the electronic structure 2, 8, 18, 7.
(a) What is the atomic number of this atom?
(b) To which of the following would it be chemically similar?
$_7N$, $_{17}Cl$, $_{15}P$, $_{18}Ar$.
(c) Why would you expect it to be similar to your answer in (b)?

4 Give a short definition of each of the following:
(a) element,
(b) compound,
(c) electron energy level,
(d) atomic number.
Use examples to explain the meaning of each.

5 Summarise the evidence which leads to each of the following true statements.
(a) All elements contain electrons.
(b) Atoms contain dense nuclei surrounded by orbiting electrons.
(c) Electrons around a nucleus can possess only certain specific energies.

CHAPTER 3

1 What sort of primary bonding would you expect between the following?
(a) Sodium and oxygen.
(b) Many atoms of magnesium.
(c) Oxygen and sulphur.
(d) Calcium and sulphur.

2 Write down the electronic structures of calcium (atomic number 20) and chlorine (atomic number 17), and show how these will react to form ionically bonded calcium chloride, $CaCl_2$.
Similarly show how magnesium nitride, Mg_3N_2, is formed.

3 Using dots and crosses for outer electrons, represent the covalent bonding in each of the following:
(a) PH_3, (b) CCl_4, (c) F_2,
(d) ClF, (e) CS_2, (f) P_2 vapour.

4   How many lone-pair electrons are held by each of the underlined atoms?
    (a) $\underline{N}H_3$,   (b) $H_2\underline{S}$,   (c) $H\underline{Cl}$.

5   Which of the following have giant structures?
    (a) Oxygen gas,   (b) iron,
    (c) copper oxide,   (d) diamond,
    (e) granite rock,   (f) carbon dioxide,
    (g) water,   (h) lighter fuel.

6   Which of the following will conduct electricity with ease?
    (a) Solid sodium chloride,
    (b) copper,
    (c) diamond,
    (d) carbon dioxide,
    (e) molten calcium chloride.
    Give a short explanation in each case for your answer.

7   Substance X has a melting point of 1500°C. It conducts electricity at room temperature.
    Substance Y is a gas at room temperature. It does not conduct electricity.
    Substance Z is soluble in water, forming a solution which does conduct electricity. However, solid Z does not conduct. Z melts at 1020°C to form a conducting liquid.
    Deduce whether the substances X, Y and Z
    (a) have giant or molecular structures,
    (b) are bonded ionically, covalently or metallically.

CHAPTER 4

1   Predict the formulae of the following compounds:
    (a) copper(II) chloride,
    (b) silver sulphite,
    (c) aluminium oxide,
    (d) fluorine oxide,
    (e) nitrogen iodide.

2   Balance the following equations.
    (a) $N_2(g) + O_2(g) \longrightarrow NO(g)$
    (b) $P(s) + Cl_2(g) \longrightarrow PCl_5(s)$
    (c) $H_2S(g) + SO_2(g) \longrightarrow H_2O(l) + S(s)$
    (d) $NaClO_3(s) \longrightarrow NaCl(s) + O_2(g)$

3   A washing up bowl holds 7200 g of water. The mass of 1 mole of water is 18 g. How many *molecules* of water will be in the bowl? Take Avogadro's number as 6×10^{23}.

4   How might the following be separated?
    (a) Water from sea water.
    (b) Coffee grounds from a mixture of these with hot water.
    (c) Sugar from a solution of sugar dissolved in water.
    (d) Different complex chemicals in a vegetable dye.
    (e) Methylated spirits from water.
Describe the method you would use in each case.

CHAPTER 5

1   Write the equations whose energy change represents
    (a) the heat of combustion of octane, C_8H_{18},
    (b) the heat of formation of copper sulphate, $CuSO_4$,
    (c) the heat of solution of sugar, $C_{12}H_{22}O_{11}$.

2   Is the combustion of petrol in a car engine an exothermic or endothermic reaction? What does your answer tell us about the energy required to break bonds in petrol and oxygen in the air compared to the energy given out on forming the combustion products?

3   Write down the equation for methane, CH_4, burning completely in oxygen gas.
    What bonds are broken and what bonds formed when this reaction occurs?

4       $N_2(g) \longrightarrow 2N(g)$ $\Delta H = +946 \text{ kJ mol}^{-1}$
        $O_2(g) \longrightarrow 2O(g)$ $\Delta H = +500 \text{ kJ mol}^{-1}$

Which of these two molecules, N_2 or O_2, contains the stronger bond? Suggest a reason for this in terms of the nature of the bonds.

5   Which of the following reactions would you expect to be exothermic, and which endothermic?

    (a) $Pb^{2+}(aq) + 2I^-(aq) \longrightarrow PbI_2(s)$

    (b) $H_2O(l) \longrightarrow H_2O(g)$

    (c) Removing two electrons from an atom.

CHAPTER 6

1 Explain why most chemical reactions become slower as they proceed.

2 Give an example from everyday life in which the rate of a reaction is altered by
  (a) changing the concentration,
  (b) changing the temperature,
  (c) dividing a solid more finely,
  (d) adding a catalyst.

3 In the equilibrium reactions below, would an increase in temperature favour the forward or the back reaction?
  (a) Heat + solid salt $\rightleftharpoons$ salt solution
  (b) $4NH_3(g) + 5O_2(g) \rightleftharpoons 4NO(g) + 6H_2O + Heat$

4 In the equilibrium reaction

$$2N_2O_5(g) \rightleftharpoons 4NO_2(g) + O_2(g)$$

how would the concentration of NO_2 be affected by an increase in pressure?

5 Describe with experimental detail how you would show that the rate of the reaction

$$2Ca(s) + 2H_2O(l) \longrightarrow 2Ca(OH)_2(aq) + H_2(g)$$

increases with temperature.

CHAPTER 7

1 What is meant by 'oxidation' and 'reduction'? What is formed when
  (a) sodium ions are reduced,
  (b) silver atoms are oxidised,
  (c) Cu^+ ions are reduced,
  (d) Cu^+ ions are oxidised?

2 In Nature, one of the most common oxidising agents is oxygen in the air. Give as many examples as you can from everyday life in which you think substances have been oxidised by the air. If the oxidation is an unwanted effect, then suggest how that reaction might be discouraged.

3   What has been oxidised and what reduced in each of the following examples?
   (a) $S(s) + O_2(g) \longrightarrow SO_2(g)$
   (b) $Ca(s) + Cl_2(g) \longrightarrow CaCl_2(s)$
   (c) $2Fe^{2+}(aq) + I_2(aq) \longrightarrow 2Fe^{3+}(aq) + 2I^-(aq)$
   (d) $Mg(s) + Cu^{2+}(aq) \longrightarrow Mg^{2+}(aq) + Cu(s)$

4   For each part of question 3, which is the oxidising agent and which the reducing agent?

CHAPTER 8

1   Where in the electrochemical series would you expect to find
   (a) a metal suitable for use in coins,
   (b) a metal suitable as a very strong reducing agent,
   (c) a metal able to react with iron oxide to produce metallic iron?
   Give reasons for each of your answers.

2   W, X, Y and Z are metals. Aqueous W chloride reacts with X to give a suspension of metallic W and X chloride solution. W will react with Y oxide to form Y and W oxide. X displaces metallic W from aqueous W nitrate, but does not displace metallic Z from aqueous Z sulphate. Arrange these four metals into the order in which they would appear in the electrochemical series.

3   Using the electrochemical series on p. 96, which of the following reactions will not occur?
   (a) $Mg(s) + Pb(NO_3)_2(aq) \longrightarrow Mg(NO_3)_2(aq) + Pb(s)$
   (b) $Zn(s) + MgO(s) \longrightarrow Mg(s) + ZnO(s)$
   (c) $Fe(s) + ZnSO_4(aq) \longrightarrow Zn(s) + FeSO_4(aq)$
   (d) $Ca(s) + 2AgNO_3(aq) \longrightarrow 2Ag(s) + Ca(NO_3)_2(aq)$

4   Describe, giving equations, what you would expect to see at each electrode when the following are electrolysed:
   (a) molten calcium chloride,
   (b) a concentrated solution of sodium bromide in water,
   (c) copper(II) nitrate solution in water,
   (d) silver nitrate solution in water.

5   Using the electrochemical series on p. 96, which of the following combinations of metals would you expect to produce the highest voltages when connected to a voltmeter (zinc connected to meter's negative terminal) and dipped into a beaker of dilute sulphuric acid?
   (a) Zinc with iron,

(b) zinc with lead,
(c) zinc with silver,
(d) zinc with copper.

CHAPTER 9

1   Suggest equations for the reaction of hydrobromic acid, HBr(aq), with
    (a) zinc,
    (b) sodium hydroxide,
    (c) calcium carbonate.
    Write the equation for hydrogen bromide gas, HBr, mixing with water.

2   Ethandioic acid, $H_2C_2O_4$, is diprotic. Explain the meaning of *diprotic*,
    and write two equations for the reaction of this acid with sodium
    hydroxide in which the normal salt and the acid salt are formed.

3   Differentiate between
    (a) a weak and a strong acid,
    (b) an alkali and a base,
    (c) a concentrated and a dilute acid.

4   Use the electrochemical series on p. 96 to determine which of the
    following metals will displace hydrogen from a dilute acid:
    (a) magnesium,
    (b) silver,
    (c) lead,
    (d) zinc.

5   Place the following aqueous solutions in order of increasing pH,
    assuming that each contains one mole of acid per cubic decimetre of
    solution:
    (a) sodium hydroxide,
    (b) ethanoic acid,
    (c) hydrochloric acid,
    (d) sulphuric acid.

6   Which of the following oxides would you expect to be basic, and
    which acidic?
    (a) Lithium oxide, Li_2O.
    (b) Sulphur dioxide, SO_2.
    (c) Nitrogen dioxide, NO_2.
    (d) Calcium oxide, CaO.
    (e) Chlorine dioxide, ClO_2.
    Give reasons for your answer.

7 Explain, with experimental details, how you would obtain a pure crystalline sample of sodium nitrate from solutions of nitric acid and sodium hydroxide of unknown concentrations.

CHAPTER 10

1 Which of the following statements are correct;
(a) Sodium atoms are larger than potassium atoms.
(b) Chlorine is more electronegative than bromine.
(c) The outer electrons of magnesium are better screened from the nucleus than those of calcium.
(d) Lithium will form positive ions more readily than beryllium.
(e) Sodium atoms are smaller than sodium ions.

2 What sort of primary bonding would you expect to find binding atoms together in the following?
(a) Zinc metal,
(b) calcium chloride,
(c) iron(II) sulphide,
(d) an alloy of magnesium and aluminium,
(e) sulphur dioxide.

3 In each of the following pairs, which ion would you expect to hydrate more readily?
(a) Li^+ or K^+,
(b) Na^+ or Mg^{2+},
(c) Mg^{2+} or Al^{3+}.
Give reasons for your answers.

4 In each of the following pairs, which compound would you expect to decompose more readily to give the oxide on heating? Why?
(a) Magnesium carbonate or calcium carbonate.
(b) Sodium hydroxide or magnesium hydroxide.
(c) Lithium nitrate or sodium nitrate.

CHAPTER 11

1 Compare the reactivity of sodium, magnesium, aluminium, iron and copper with
(a) oxygen gas,
(b) water,
(c) dilute hydrochloric acid.
Give equations and name the products.

2 Compare the behaviour of calcium oxide, aluminium oxide and copper oxide with
   (a) water,
   (b) dilute hydrochloric acid,
   (c) dilute sodium hydroxide.

3 How would you accomplish the following?
   (a) Convert copper metal into copper(II) ions.
   (b) Differentiate between samples of potassium chloride and sodium chloride.
   (c) Improve the corrosion resistance of aluminium.
   (d) Obtain a sample of hydrogen from aqueous sodium hydroxide.
   (e) Differentiate between two solutions, one containing iron(II) ions and the other iron(III) ions.
   (f) Obtain pure copper metal from a sample of copper(II) sulphate.

4 Compare the reduction of aluminium ore with that of iron ore. State three advantages that aluminium has over iron as an industrial metal, explaining the chemical reasons behind each advantage. What are the chief disadvantages of aluminium?

5 From which group of the Periodic Table might each of the following metals come?

   (a) X is a dense, tough metal. It has chlorides with the formulae XCl_2 and XCl_3 which are blue and green respectively.
   (b) Y is a metal which tarnishes rapidly in air. With water it reacts steadily to form a cloudy suspension. After filtering, the solution is found to be weakly alkaline. This solution forms a milky white precipitate when carbon dioxide is bubbled through.
   (c) Z is a metal whose carbonate decomposes on heating to form Z oxide with formula ZO. Z burns with a bright flame, and when dropped into water it sinks, reacting to produce bubbles of hydrogen only slowly.
   (d) W is a soft metal. W nitrate melts on heating and then gives off only oxygen gas.

6 Write an essay linking any three properties of metals to their uses for mankind. Explain the properties you have chosen in relation to chemical structure and bonding.

312

CHAPTER 12

1   Which of the following statements are correct?
    (a) Bromine is more electronegative than iodine.
    (b) Chlorine and fluorine atoms in the compound ClF are bound
        covalently.
    (c) The oxygen atom is smaller than the fluorine atom.
    (d) When sulphur forms the sulphide ion it assumes the electronic
        structure of argon.
    (e) All non-metals are non-conductors of electricity.
    (f) Sulphur is a better oxidising agent than chlorine.

2   In terms of primary and secondary bonding, explain why most metallic
    oxides are non-volatile solids when most non-metallic oxides are
    volatile. Why is silicon dioxide an exception to this rule?

3   Write equations for two reactions in which elements from group 7 of
    the Periodic Table form covalent bonds and two in which elements
    from this group form ionic bonds. Compare the physical properties of
    the products and explain their differences.

4   Which of the following reactions will not occur? Why not?
    (a) $2Cl^-(aq) + I_2(aq) \longrightarrow Cl_2(aq) + 2I^-(aq)$
    (b) $2I^-(aq) + Br_2(aq) \longrightarrow I_2(aq) + 2Br^-(aq)$
    (c) $2Cl^-(aq) + Br_2(aq) \longrightarrow Cl_2(aq) + 2Br^-(aq)$

5   How would you accomplish the following?
    (a) Chemically lower the water vapour content of a room.
    (b) Demonstrate that breath contains water vapour.
    (c) Remove oxygen from a mixture of oxygen and nitrogen.
    (d) Show that the gas coming from a reaction in a test tube was
        hydrogen.
    (e) Determine the volume of carbon dioxide present in a flask
        containing carbon dioxide and nitrogen.

6   Outline how you might convert a small sample of sulphur into sul-
    phurous acid, and thence to sodium sulphite crystals.

7   Show by means of equations, giving essential reaction conditions, how
    concentrated sulphuric acid reacts with
    (a) sodium chloride,
    (b) water.
    Similarly show how dilute sulphuric acid reacts with
    (a) sodium hydroxide,

(b) zinc,
(c) sodium carbonate,
(d) universal indicator.

8   Show how the chemical principles involved in the contact process and the Haber process influence the choice of conditions in these industrial reactions.

9   Devise an experiment to demonstrate that the formula of water is H_2O.

10 Identify the following sodium salts.
    (a) *W* dissolves in water, but addition of hydrochloric acid followed by barium chloride solution produces a white precipitate.
    (b) A solution of *X* in water, after acidifying with nitric acid followed by the addition of silver nitrate solution, forms a yellow precipitate. When chlorine gas is bubbled through a separate solution of *X* in water an intense brown colour is given.
    (c) *Y* is a solid which fizzes on contact with dilute sulphuric acid. If the gas that is given off is bubbled through lime water a white cloudiness is produced.
    (d) *Z* is a solid which also fizzes when dilute sulphuric acid is added. In this case a gas with a choking smell is given off which turns moist potassium dichromate paper from orange to green, and moist universal indicator paper red.

11 One liquid household cleaner, *A*, turns universal indicator solution blue. It has a very strong smell, and when held near an unstoppered bottle of hydrochloric acid dense white fumes form. A second household cleaner, *B*, is slightly acidic. It also has a strong smell, and when held near moist universal indicator paper the paper turns white.
    Identify the gases coming from *A* and *B*.

CHAPTER 13

1   Draw out and name
    (a) the two isomers of propanol, C_3H_7OH,
    (b) the three isomers of butene, C_3H_6.

2   Draw out the structures of
    (a) 2,2-dimethylbutane,
    (b) propanoic acid,
    (c) methylmethanoate.

3   Give one equation and the essential conditions in each case which illustrate the following:
    (a) a substitution reaction,
    (b) an addition reaction,
    (c) hydrolysis,
    (d) the incomplete combustion of an alkane.

4   Describe a chemical test to differentiate between
    (a) pentene and pentane,
    (b) ethanol and ethanoic acid.

5   Give an outline of the experiments you would carry out in order to obtain
    (a) propanoic acid from propanol,
    (b) ethylethanoate from ethanol and ethanoic acid.

6   Explain the meaning of the terms
    (a) reflux,
    (b) unsaturated,
    (c) cracking,
    (d) functional group,
    (e) homologous series.

7   'The importance of crude oil is not only as a source of petrol.' Write an essay in support of this statement.

CHAPTER 14

1   Use one example in each case to explain the following terms:
    (a) condensation polymerisation,
    (b) addition polymerisation,
    (c) hydrolysis of a polymer,
    (d) cross-linking,
    (e) enzyme.

2   Which of the following would definitely make poor detergents? Why?
    (a) $CH_3CH_2CH_2CH_2CH_2CH_3$,
    (b) $CH_3CH_2CH_2CH_2CH_2OPO_3Na_2$,
    (c) $CH_3CH_2CH_2CH_2CH_2CH_2ONa$,
    (d) Na_2SO_4.

3   Show how you might expect the following to polymerise.
    (a) $HOCH_2CH_2CH_2OH$ with $HOOCCH_2CH_2CH_2CH_2COOH$,

(b) $HOCH_2CH{=}CH_2$,

(c) $HOCH_2CH_2CH_2COOH$.

4   Devise an experiment to compare the effectiveness of three different enzymes in breaking down starch at body temperature in neutral solution.

CHAPTER 15

(Use the table of relative atomic masses on p.262.)

1   What is the mass of each of the following?
(a) One mole of hydrogen atoms.
(b) One mole of hydrogen molecules.
(c) One mole of carbon monoxide molecules.
(d) 0.04 moles of sodium chloride.
(e) 0.5 moles of hydrated copper(II) sulphate, $CuSO_4.5H_2O$.

2   How many atoms are present in 0.02 moles of helium gas?

3   (a) How many moles of atoms are present in 1.15 g of sodium metal?
(b) How many moles of molecules are present in 1 kg of water?
(c) How many molecules and how many atoms are present in 23 g of ethanol, C_2H_5OH?

4   What mass of carbon dioxide would be lost when excess sulphuric acid is added to 200 g of copper(II) carbonate?

$$CuCO_3(s) + H_2SO_4(aq) \longrightarrow CuSO_4(aq) + H_2O(l) + CO_2(g)$$

5   What mass of silver is contained in 100 g of silver nitrate, $AgNO_3$?

6   What mass of carbon would be needed to produce 1 kg of carbon monoxide?

$$2C(s) + O_2(g) \longrightarrow 2CO(g)$$

What volume, at room temperature and pressure, would this gas occupy?

7   What volume of oxygen, at room temperature and pressure, would be used up for every gram of liquid octane burned by an engine?

$$C_8H_{18}(l) + 12\tfrac{1}{2}O_2(g) \longrightarrow 8CO_2(g) + 9H_2O(l)$$

8 A balloon containing 2000 dm³ of helium gas is released from ground level at a temperature of 20°C and at 1 atmosphere pressure. What volume will the helium occupy at a height where the pressure is only 0.5 atmospheres and the temperature 5°C?

9 Calculate the density of oxygen gas in grammes per cubic centimetre at room temperature and pressure, taking the relative molecular mass of oxygen as 32.

CHAPTER 16

(Use the table of relative atomic masses on p. 262)

1 Determine the empirical formulae of the following compounds:
   (a) contains 0.84 g carbon combined with 1.12 g oxygen only,
   (b) containing 22.3% magnesium, 33.0% chlorine and 44.7% oxygen,
   (c) contains 0.414 g sodium, 0.252 g nitrogen and 0.576 g oxygen only,
   (d) contains 0.36 g carbon and 3.19 g chlorine only.
For the compound in (d) suggest a possible structural formula.

2 What is the degree of hydration in hydrated lithium chloride, $LiCl.xH_2O$, given that 4.45 g of the hydrated salt contain 3.18 g of water.

3 What is the molarity of each of the following?
   (a) A solution containing 4.0 g sodium hydroxide, NaOH, in 1 dm³.
   (b) A solution containing 1.05 g nitric acid, HNO_3, in 50 cm³.
   (c) A solution of sodium hydroxide, 25.0 cm³ of which requires 20 cm³ of 2 molar sulphuric acid for complete neutralisation.

4 20.0 cm³ of a 0.02 molar solution of potassium manganate(VII), $KMnO_4$, were found to react with 40.0 cm³ of a 0.05 molar solution of iron(II) sulphate. Determine the number of moles of iron(II) sulphate that react with 1 mole of potassium manganate(VII).

5 Determine the relative molecular mass of the following gases:
   (a) 0.20 g of gas X occupied 282 cm³ at room temperature and pressure,
   (b) 0.79 g of Y occupied 4.47 dm³ at room temperature and pressure.

6 Calculate the mass of silver which would be deposited during silver plating if a current of 0.02 amp was used for 1 hour. What mass of copper would be plated by this same quantity of electricity?

7 Write equations for the reactions at the cathode and at the anode when dilute sulphuric acid is electrolysed

Calculate the current that would be required to produce 1 dm^3 of oxygen gas from the anode every hour.

8 The relative atomic mass of a certain metal is 59. A current of 4.0 amps deposits 4.4 g of this element every hour during the electrolysis of its chloride. Determine the charge on the metal ion.

9 5.3 g of anhydrous sodium carbonate were added to 100 cm^3 water in a polystyrene beaker. A temperature rise of 2°C was recorded. Calculate the heat of solution of sodium carbonate. What are the main sources of error in this type of determination?
(Take the specific heat of water as 4.2 J g^{-1} °C^{-1}.)

10 Describe how you would carry out an experiment to determine the heat of combustion of butane, using a small cylinder of the compressed gas. The heat of combustion of butane, C_4H_{10}, is -2900 kJ mol^{-1}. Assuming that 50% of the heat is lost, determine the mass of butane needed to boil 1 kg water if its initial temperature was 20°C.

INDEX

Reduction 86
of metal ores 166
Refluxing 231
Relative atomic mass 261
values 262
Relative molecular mass 264
determination for gases 288
Reversible reaction 78
Rubber 252
Rubidium 140
Rusting 160, 161

S
Sacrificial zinc 109
Safety 47
Salt 119
Saturation 227
Screening electrons 133
Secondary bonding 37
Shapes of molecules 28
Silicon 200
Silicon dioxide 204
Slag 169
Soap
action of 244
manufacture of 242
Sodium 140
industrial manufacture of 167
Sodium aluminate 155
Sodium amalgam 105, 171
Sodium chloride
bonding 21
preparation 122
Sodium dihydrogen phosphate 123
Sodium hydrogen sulphate 123
Sodium hydroxide
manufacture 171
Sodium peroxide 144, 145
Sodium phosphate 123
Sodium sulphate
preparation 122
Sodium vapour lamp 13, 147
Specific heat 298
Stability 21
Starch 253
Stearic acid 241
Steel manufacture 171
Strong acids 126
Strontium 148
Substitution reaction 224
Sulphates
test for 192

Sulphites
test for 191
Sulphur 188
oxides and oxoacids 191
Sulphur dioxide
test for 191
Sulphur trioxide 192
equilibrium of 92
Sulphuric acid 192
commercial manufacture 213
Sulphurous acid 191
Surface area
reaction rate 77

T
Tarnishing
of group 1 metals 145
of group 2 metals 150
Techniques 41
Teflon 247
Temperature
reaction rate 73
Temporary hardness 208
Terylene 247
Tetrahedral angle 30
Tetrahedral structure 30
of alkanes 219
Thermochemistry
calculations from 297
Titration 120
calculations from 285
Transition elements 157
chemical reactions 159
general properties 159
position in Periodic Table 158

U
Ultra-violet light 224
Universal indicator 120, 124
Unsaturation 227
Urea 217

V
Valency 217
Vanadium pentoxide as catalyst 76
Volatile 39
Voltameter 99
Volume of gases 270
Vulcanisation 252